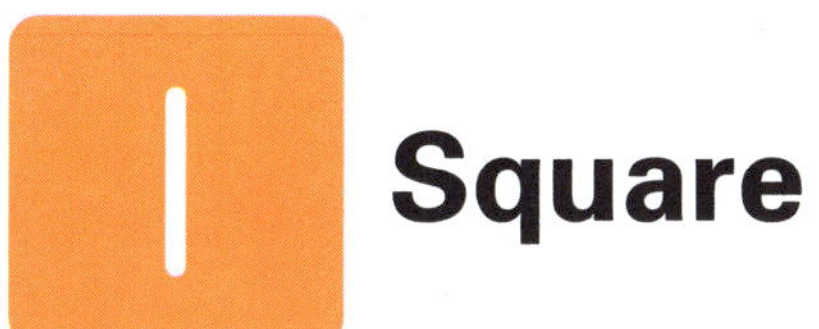

1 Square

Name

Date

10797519

To parents
Please have your child work with crayons throughout this workbook so that he or she can become familiar with not only shapes but colors as well. Your child can use any color he or she likes for this activity.

■ Draw a line from the dot (●) to the star (★).

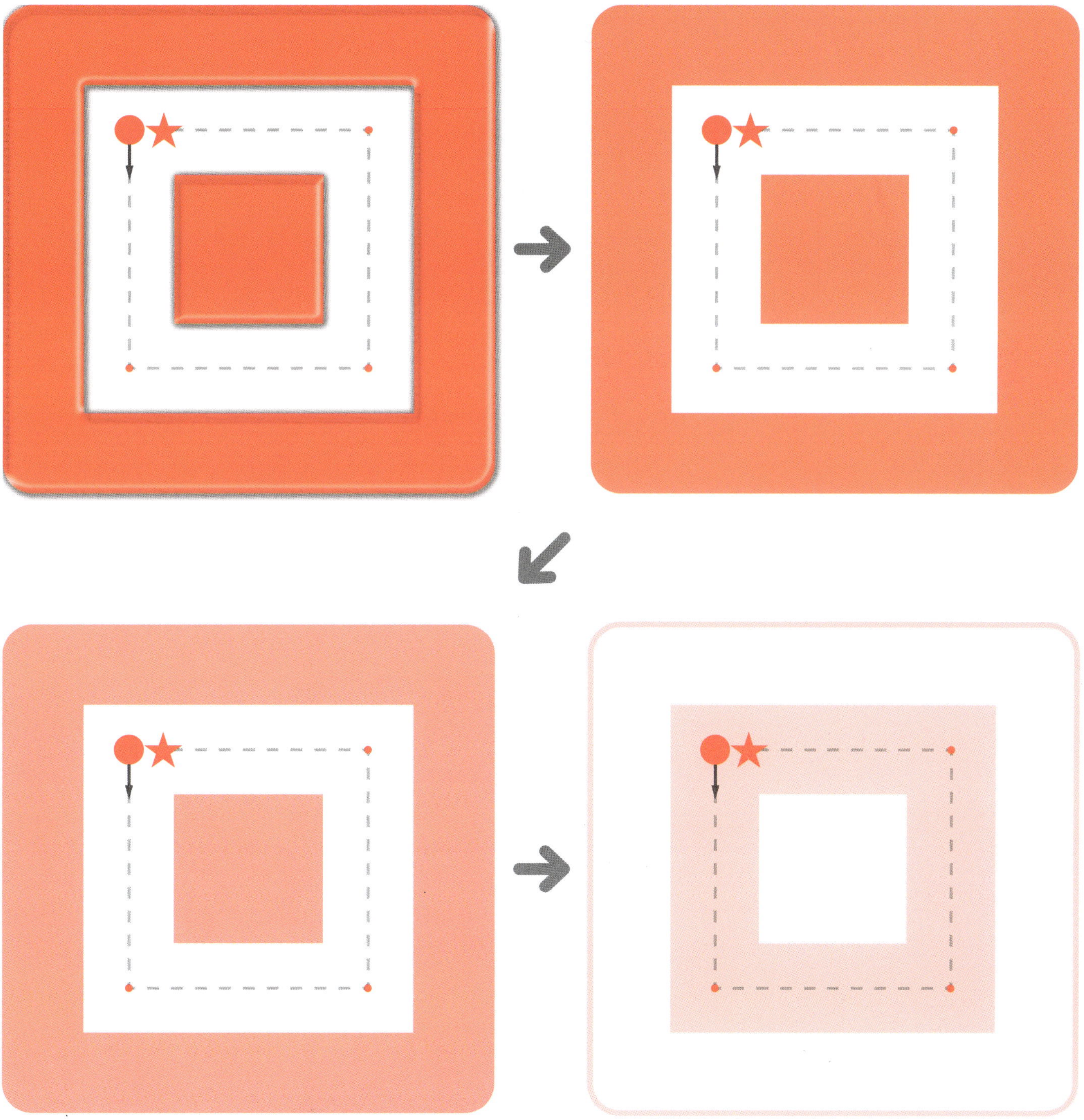

To parents
In these exercises, your child will learn the names and the characteristics of common shapes while refining his or her motor control skills. If he or she draws outside the path, help guide your child's hand and draw together to demonstrate.

■ Draw a line from the dot (●) to the star (★).

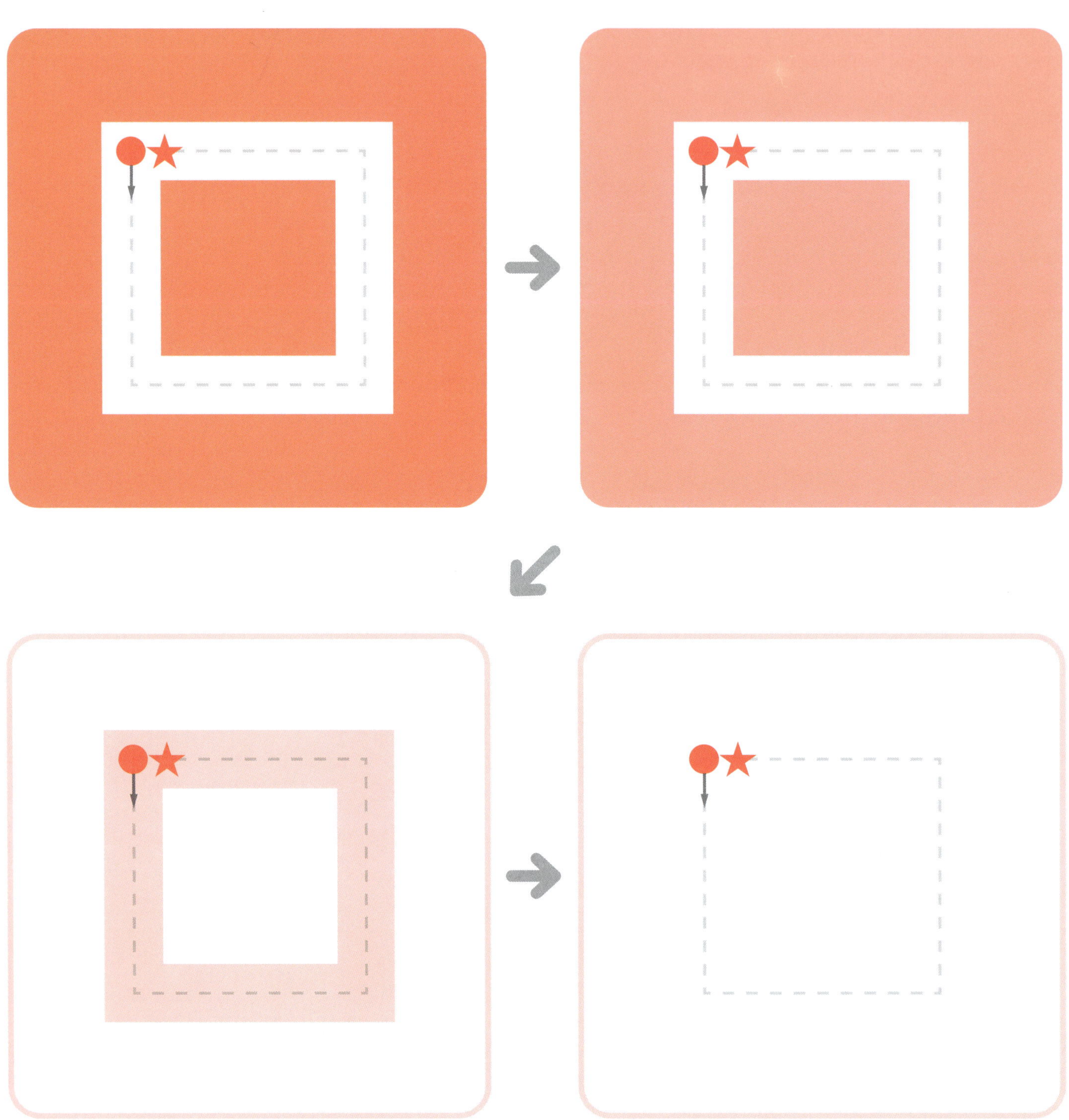

2 Squares: House

Name

Date

To parents
Each illustration includes multiple examples of the shapes that make the objects. Your child will also learn the colors to be used in each activity. It is good to say the name of the shapes and the objects, as well as the colors aloud with your child.

■ Draw a line from one crayon to the matching crayon.

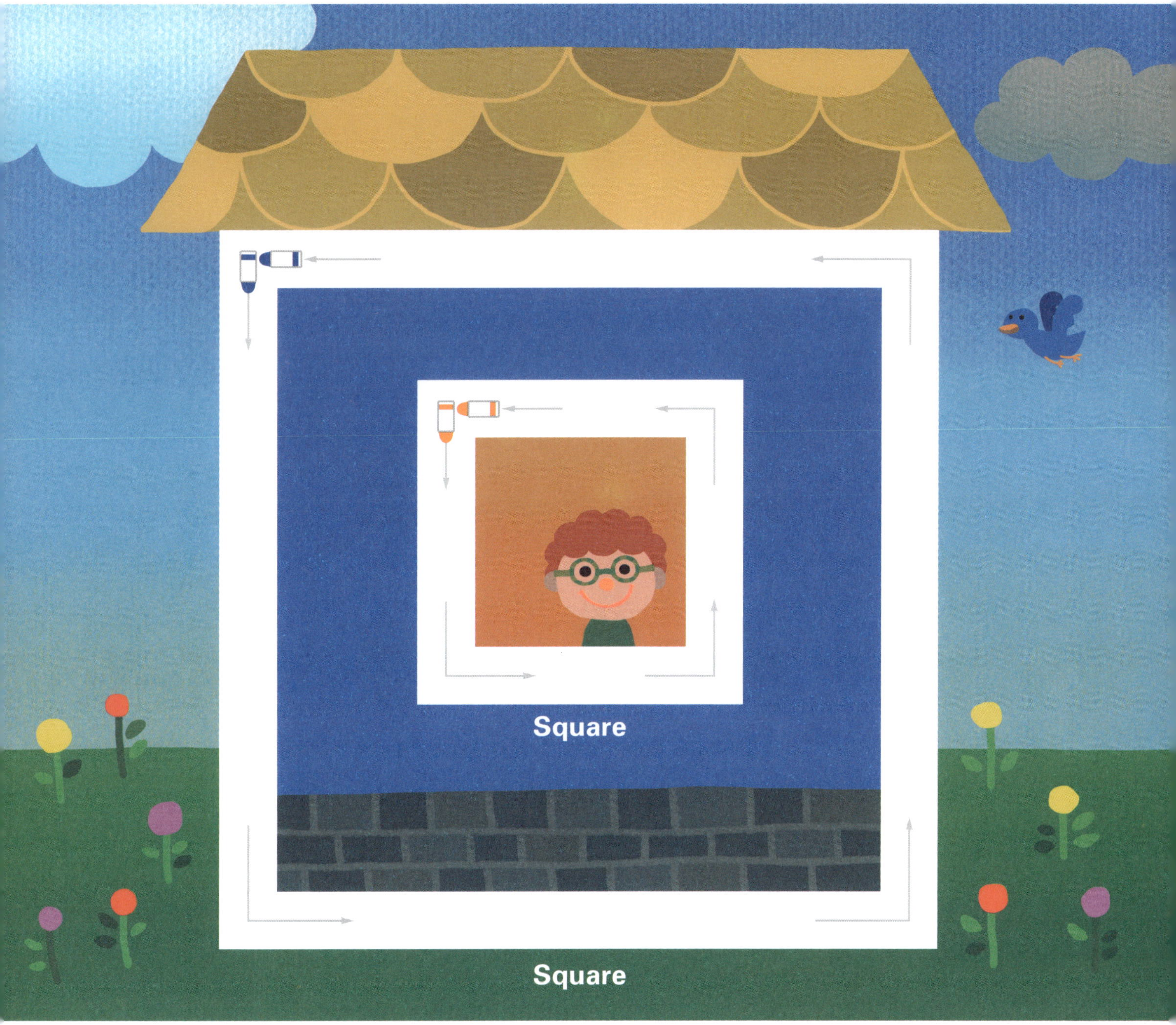

blue orange

Square

Square

To parents
If your child is having difficulty distinguishing shapes, please help him or her with the maze activity. Encourage your child to say the name of each shape and its color aloud while following the maze path.

■ Draw a line from the arrow (↓) to the star (★) by connecting the green squares (■).

3 Rectangle

Name

Date

To parents
It is okay if your child draws wavy lines at first: his or her motor control skills will improve with practice. However, it is important to have your child keep in mind the characteristics of the shape. For example, a rectangle has four right angles.

■ Draw a line from the dot (●) to the star (★).

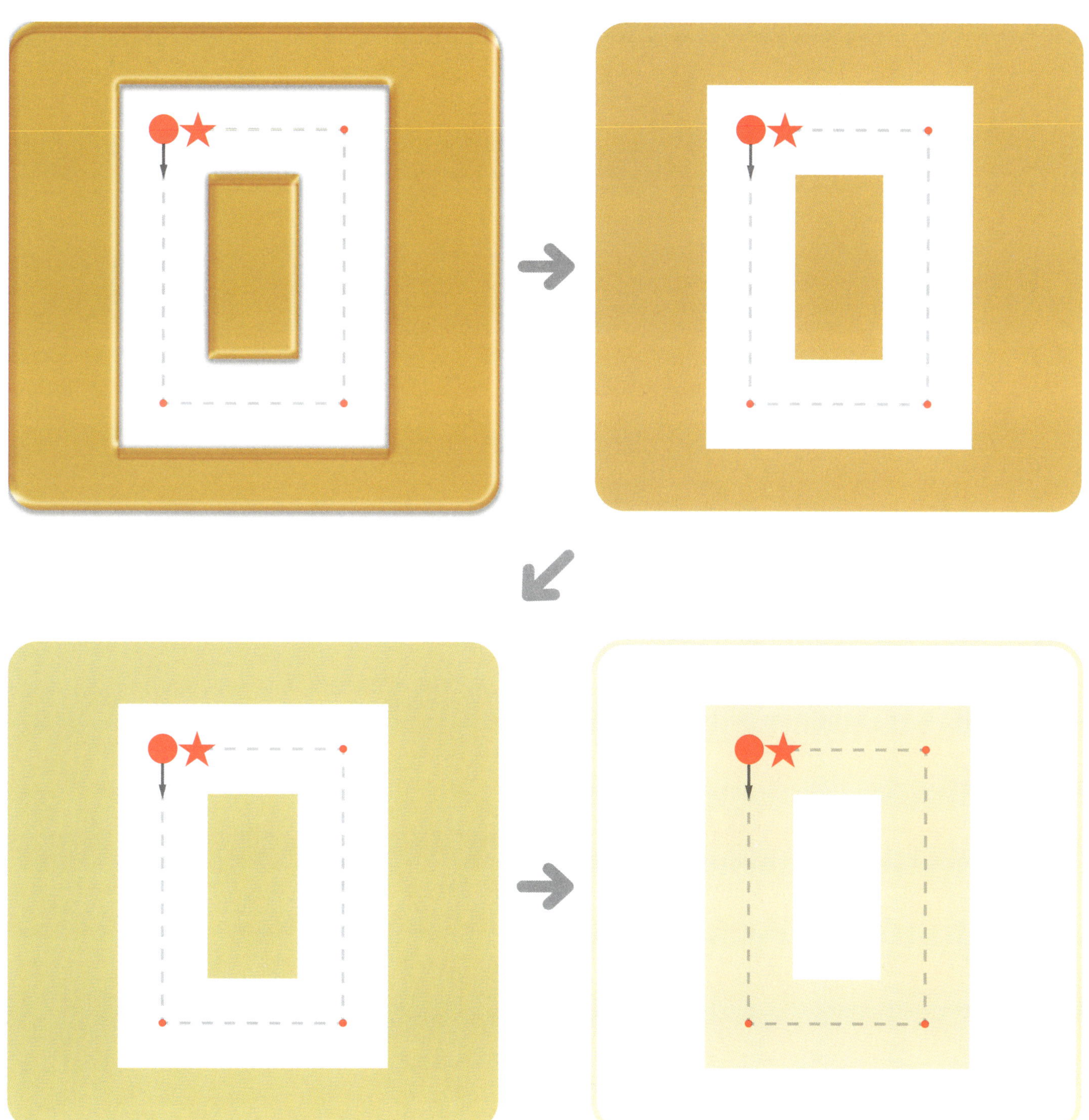

■ Draw a line from the dot (●) to the star (★).

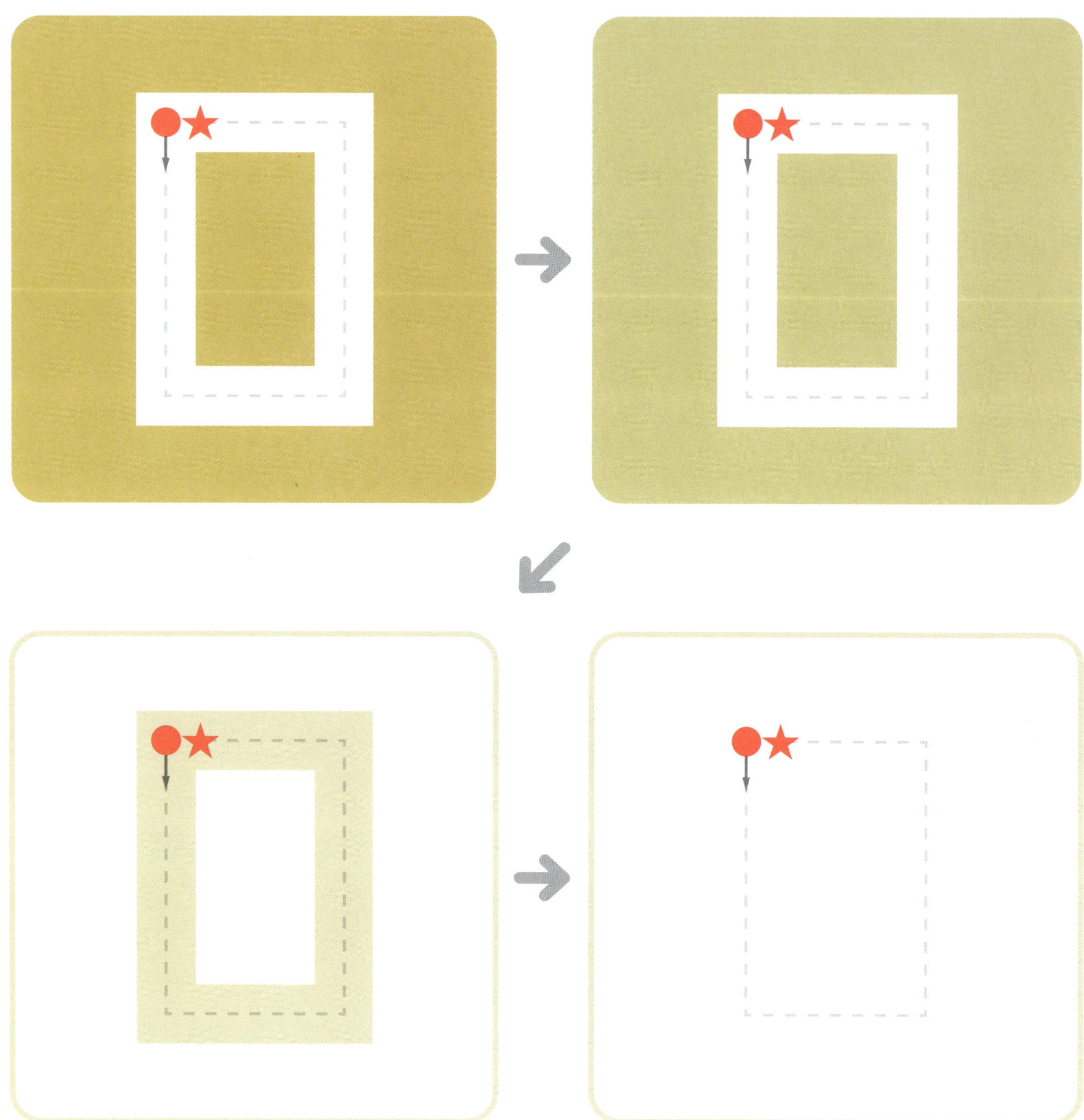

4 Rectangles: City Streets

Name

Date

To parents
When your child has finished this exercise, talk with him or her about the objects they drew. You can also talk about other rectangle shaped objects in your everyday life like bags, books, and so on.

■ Draw a line from one crayon to the matching crayon.

black brown

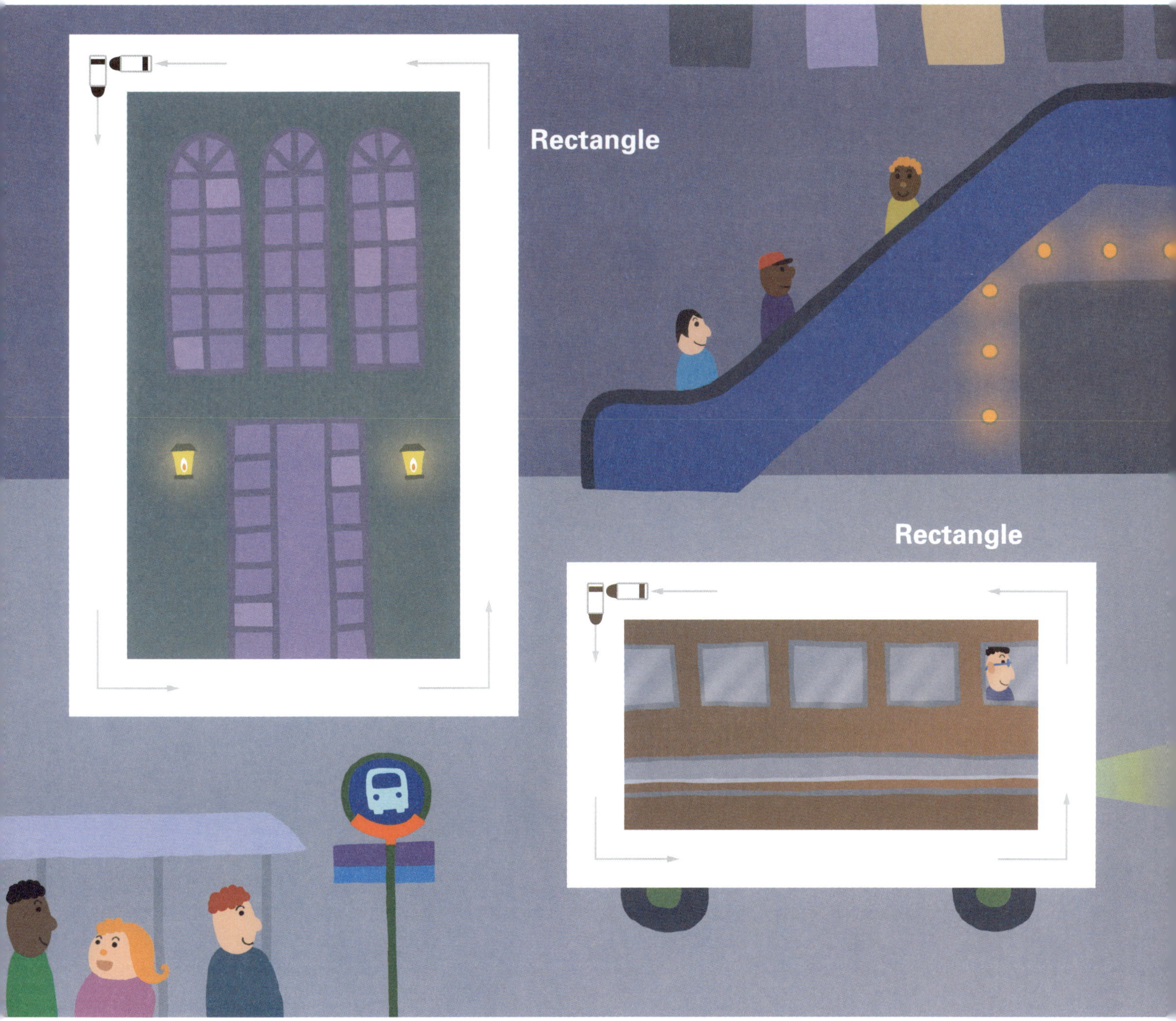

Rectangle

Rectangle

■ Draw a line from the arrow (⬇) to the star (★) by connecting the yellow rectangles (▮).

5 Triangle

Name

Date

To parents

It is okay if your child cannot draw the shape in one stroke at first. The important thing is that he or she can learn the characteristics of each shape. When your child has finished the exercise, offer him or her a lot of praise.

■ Draw a line from the dot (●) to the star (★).

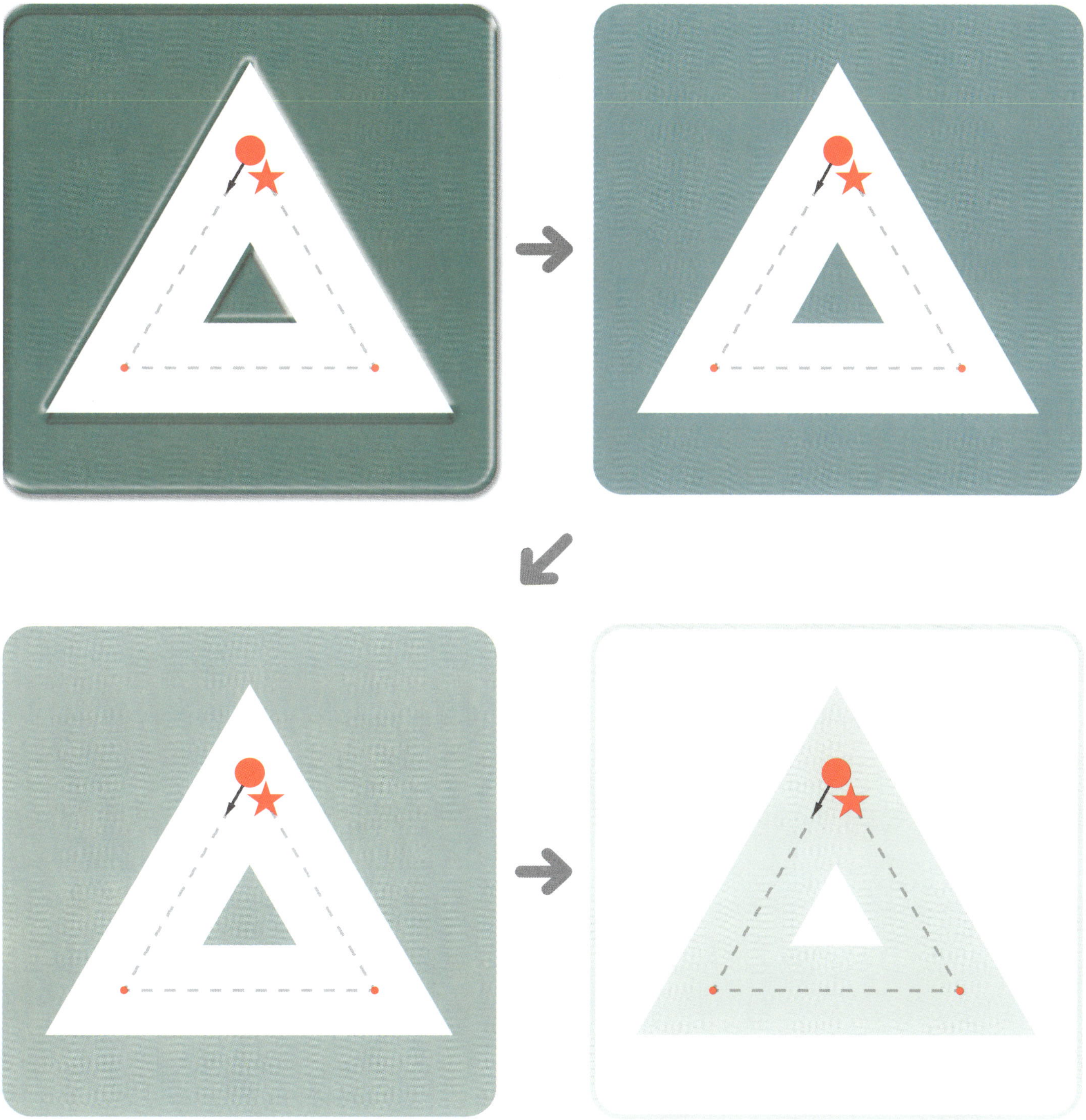

Draw a line from the dot (●) to the star (★).

6 Triangles: Tents

Name

Date

To parents
When your child has finished this exercise, talk with him or her about the objects and the colors. For example, you can say "Which tent do you like, the small violet (purple) one or the big green one?"

■ Draw a line from one crayon to the matching crayon.

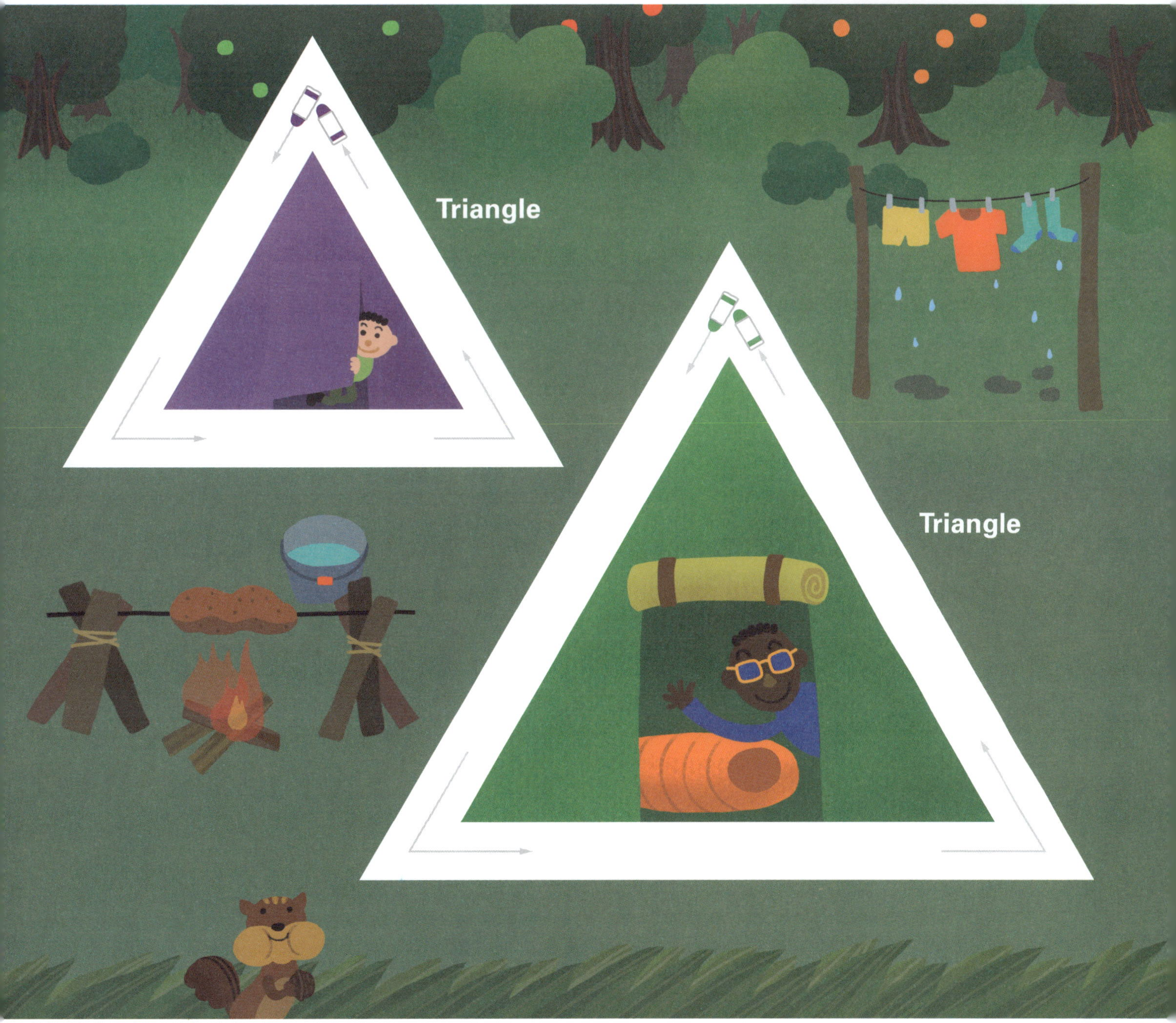

Triangle

Triangle

■ Draw a line from the arrow (⬇) to the star (★) by connecting the brown triangles (▲).

Diamond (Rhombus)

Name

Date

To parents
If your child is having difficulty drawing the shape, please remind him or her to draw slowly and carefully. It might be good to ask your child what each shape looks like to him or her.

■ Draw a line from the dot (●) to the star (★).

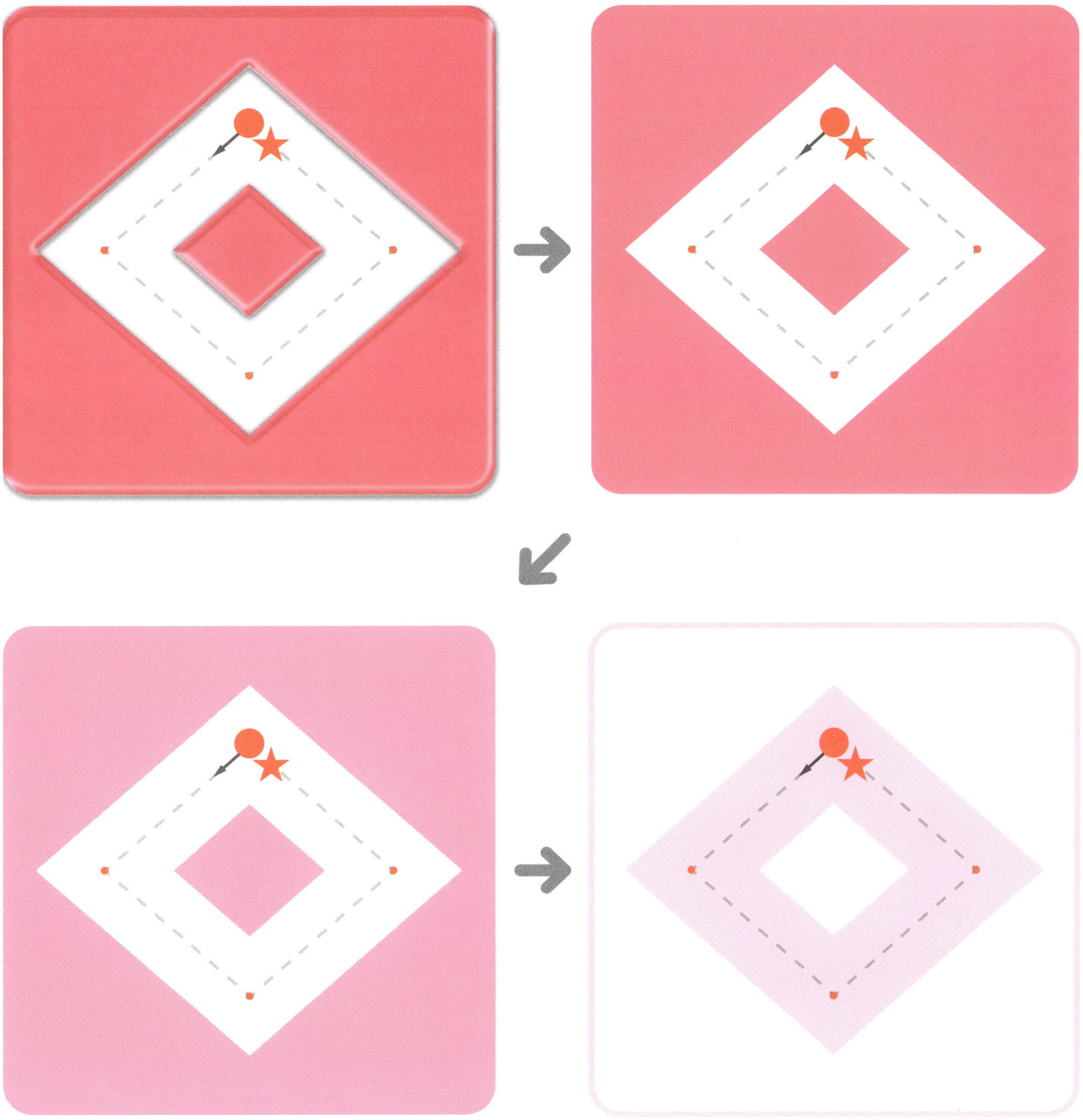

■ Draw a line from the dot (●) to the star (★).

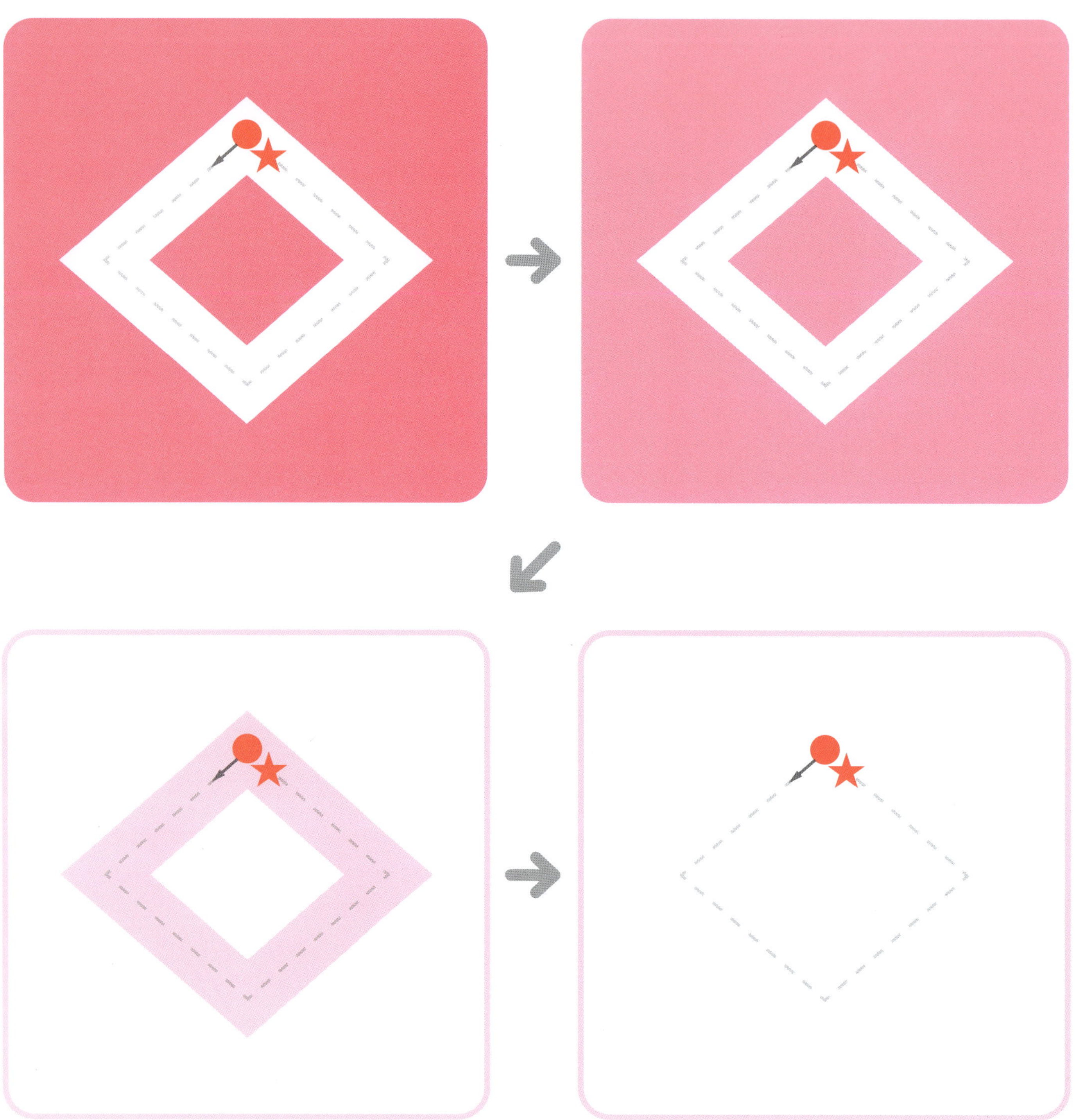

Diamonds (Rhombuses): Angel Fish

Name

Date

To parents
When your child has finished this exercise, talk with him or her about the objects and the colors. For example, you can say "Which angel fish do you like? Can you find the difference between them?"

■ Draw a line from one crayon to the matching crayon.

Diamond (Rhombus)

Diamond (Rhombus)

■ Draw a line from the arrow (↓) to the star (★) by connecting the red diamonds(rhombuses) (◆).

9 Circle

Name

Date

To parents
Have your child draw the shape while saying the name aloud at the same time. It might be a good way for him or her to learn the shape. When your child can draw a beautiful circle, offer him or her a lot of praise.

■ Draw a line from the dot (●) to the star (★).

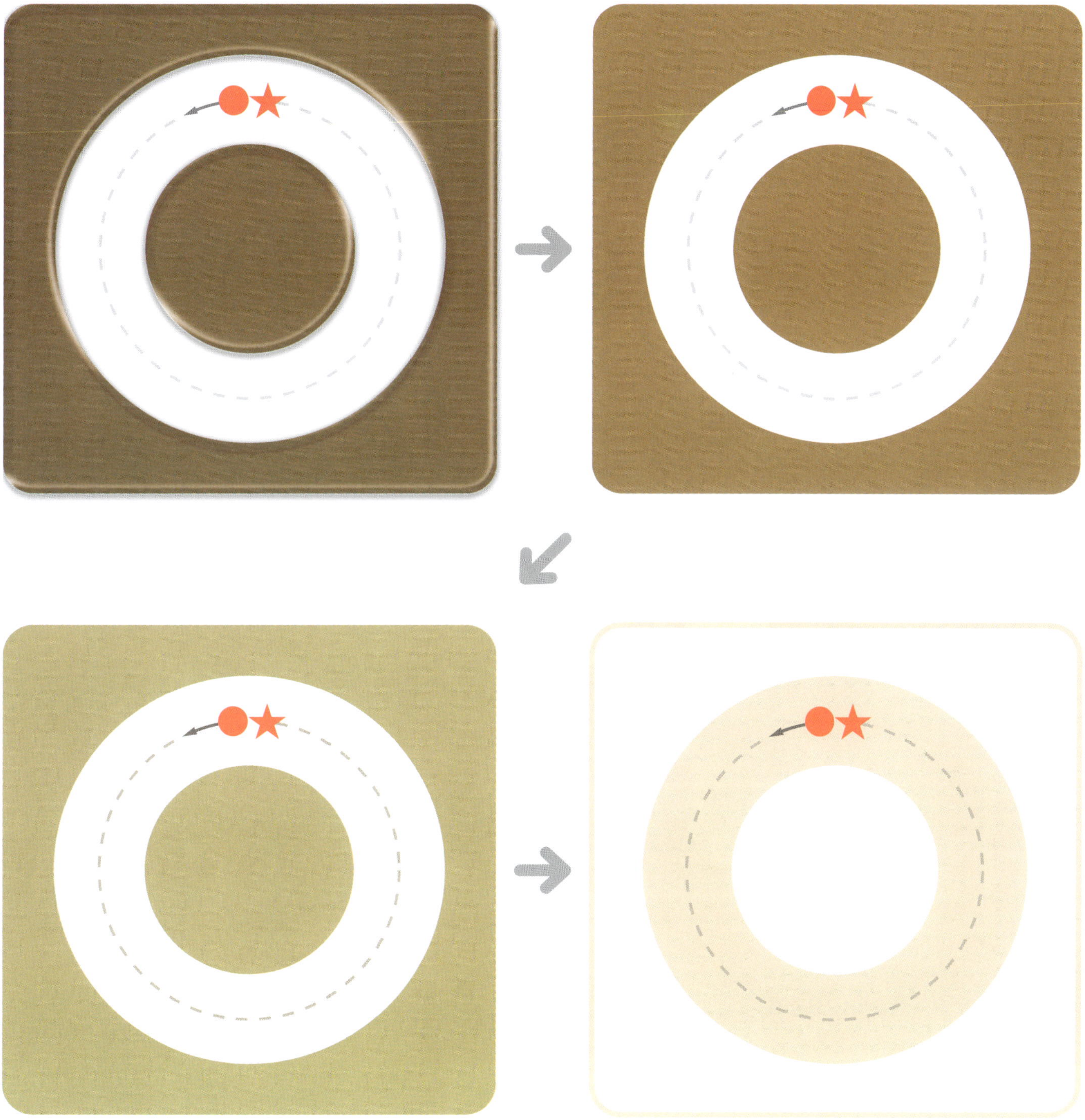

Draw a line from the dot (●) to the star (★).

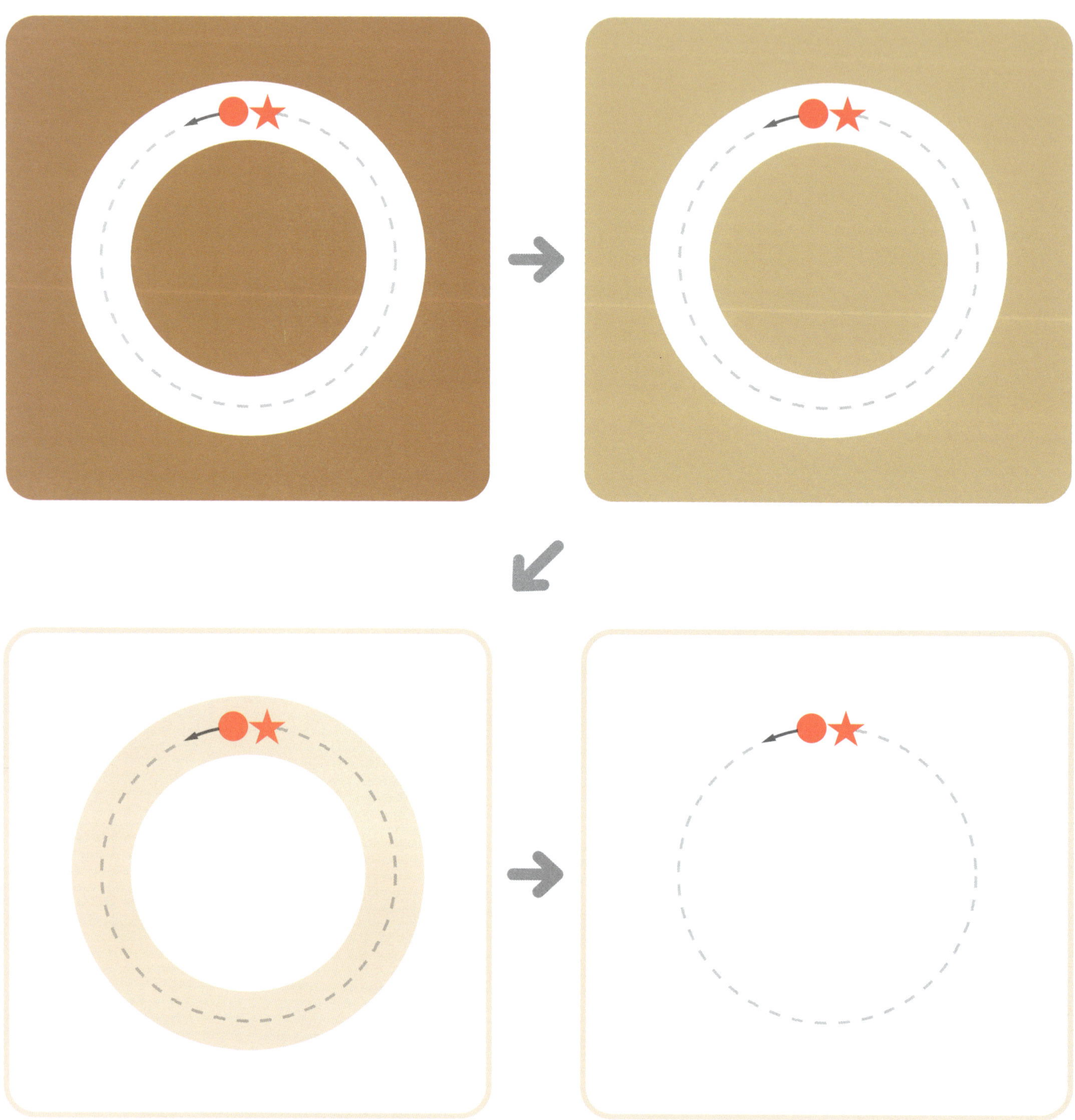

10 Circles: Bicycle Wheels

Name

Date

To parents
When your child has finished this exercise, talk with him or her about the objects they drew. You can also talk about circle shaped objects in your everyday life like balls, buttons, and so on.

■ Draw a line from one crayon to the matching crayon.

red green

Circle

Circle

■ Draw a line from the arrow (↓) to the star (★) by connecting the orange circles (●).

Oval

Name

Date

To parents
If your child is having difficulty drawing the shape, please remind him or her to draw slowly and carefully. It might be good to ask your child, what does this shape look like?

■ Draw a line from the dot (●) to the star (★).

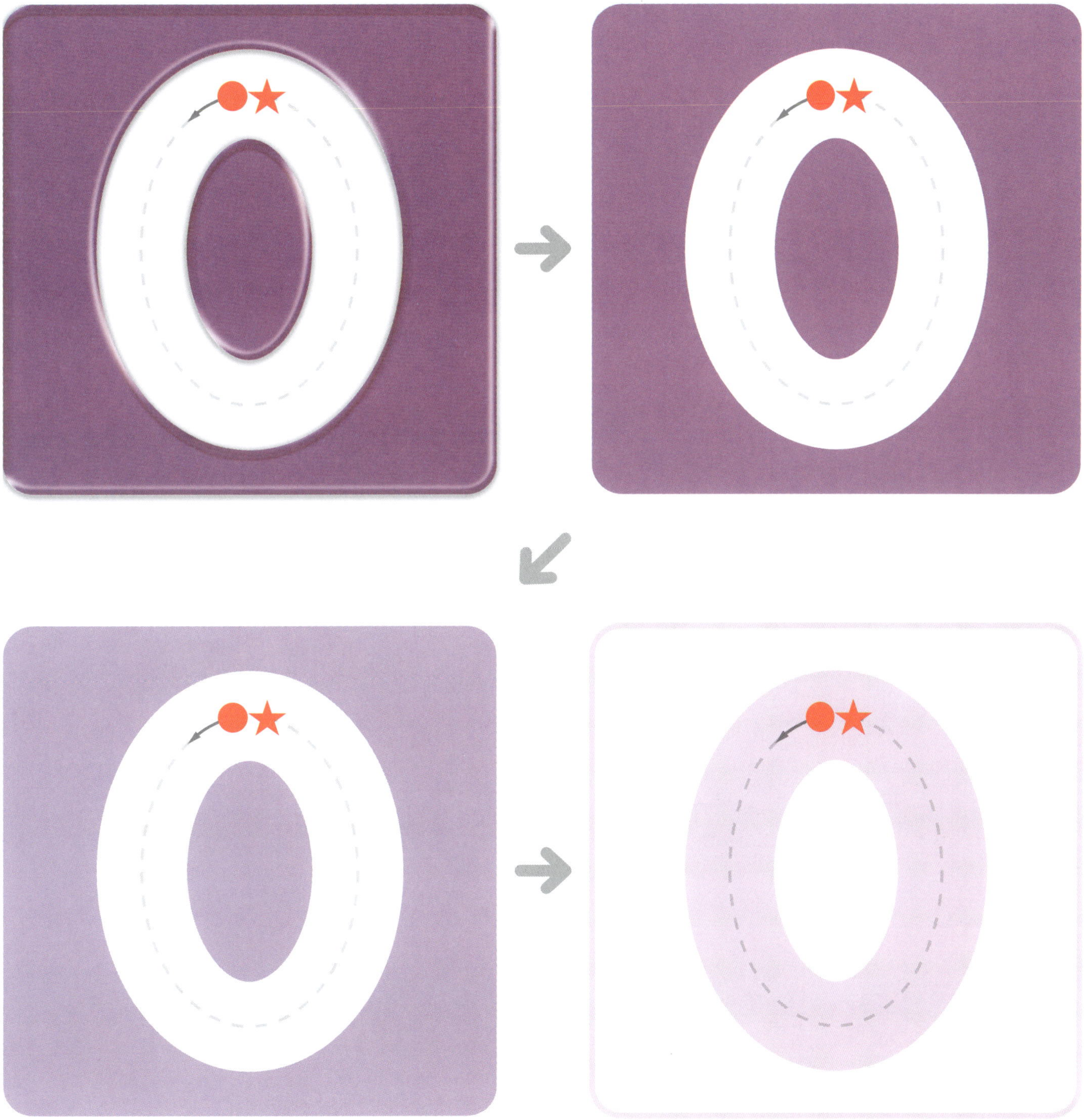

Draw a line from the dot (●) to the star (★).

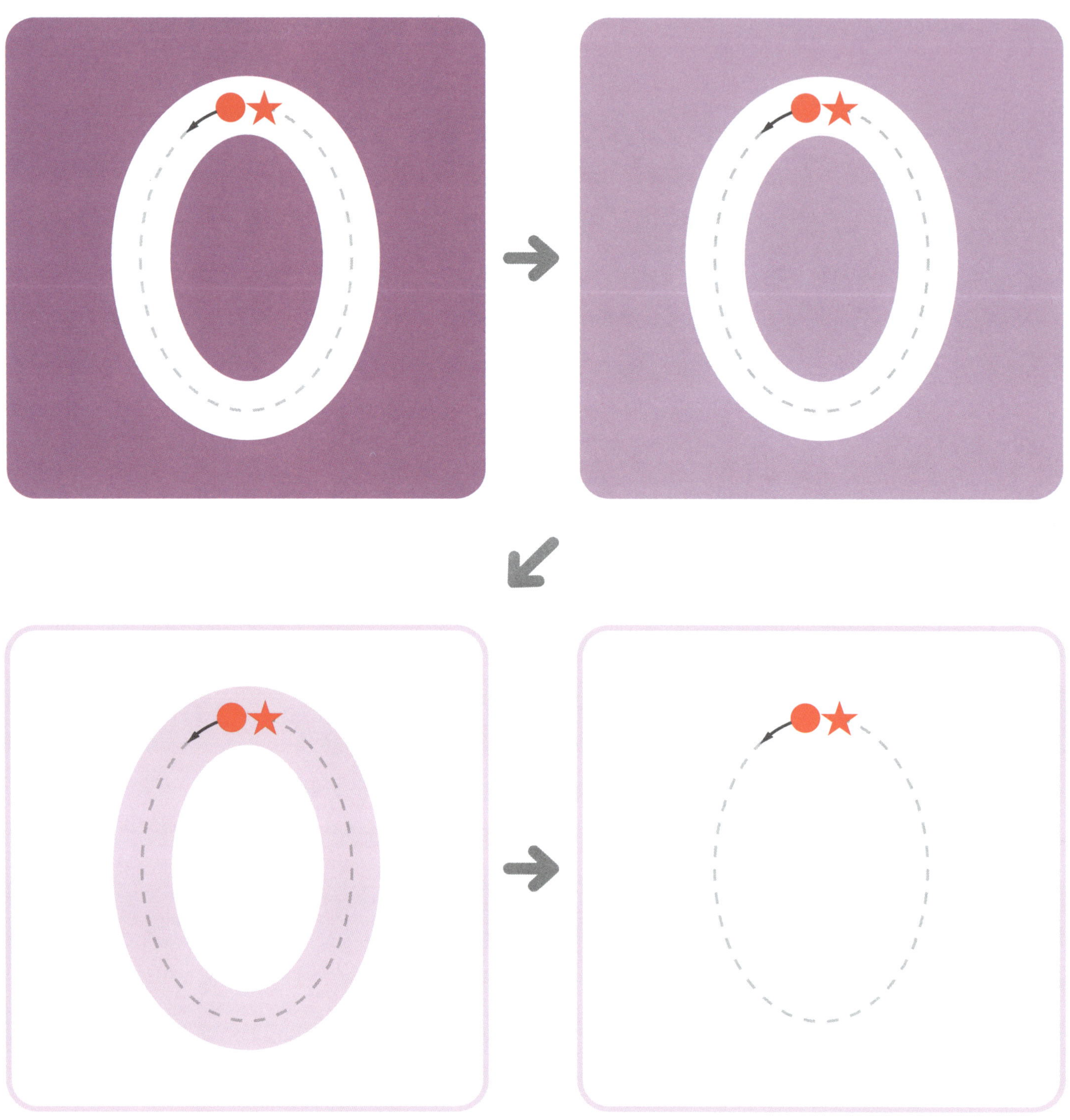

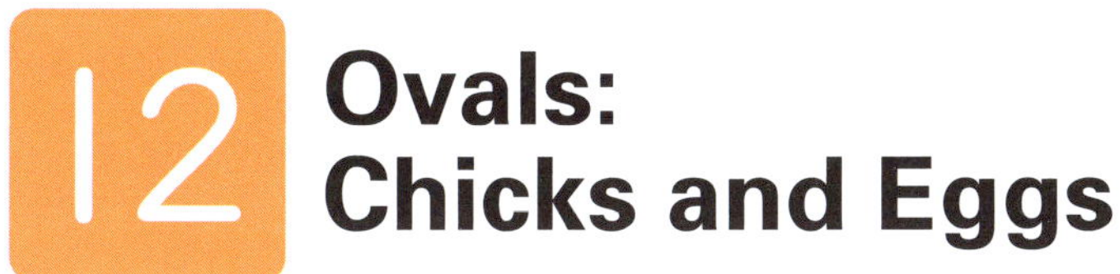

12 Ovals: Chicks and Eggs

Name

Date

To parents

The illustration shows two types of ovals: vertical (chick) and horizontal (egg.) If your child can recognize that both are the same shape in spite of the different orientation, offer him or her a lot of praise.

■ Draw a line from one crayon to the matching crayon.

Oval

Oval

■ Draw a line from the arrow (⬇) to the star (★) by connecting the blue ovals (⬮).

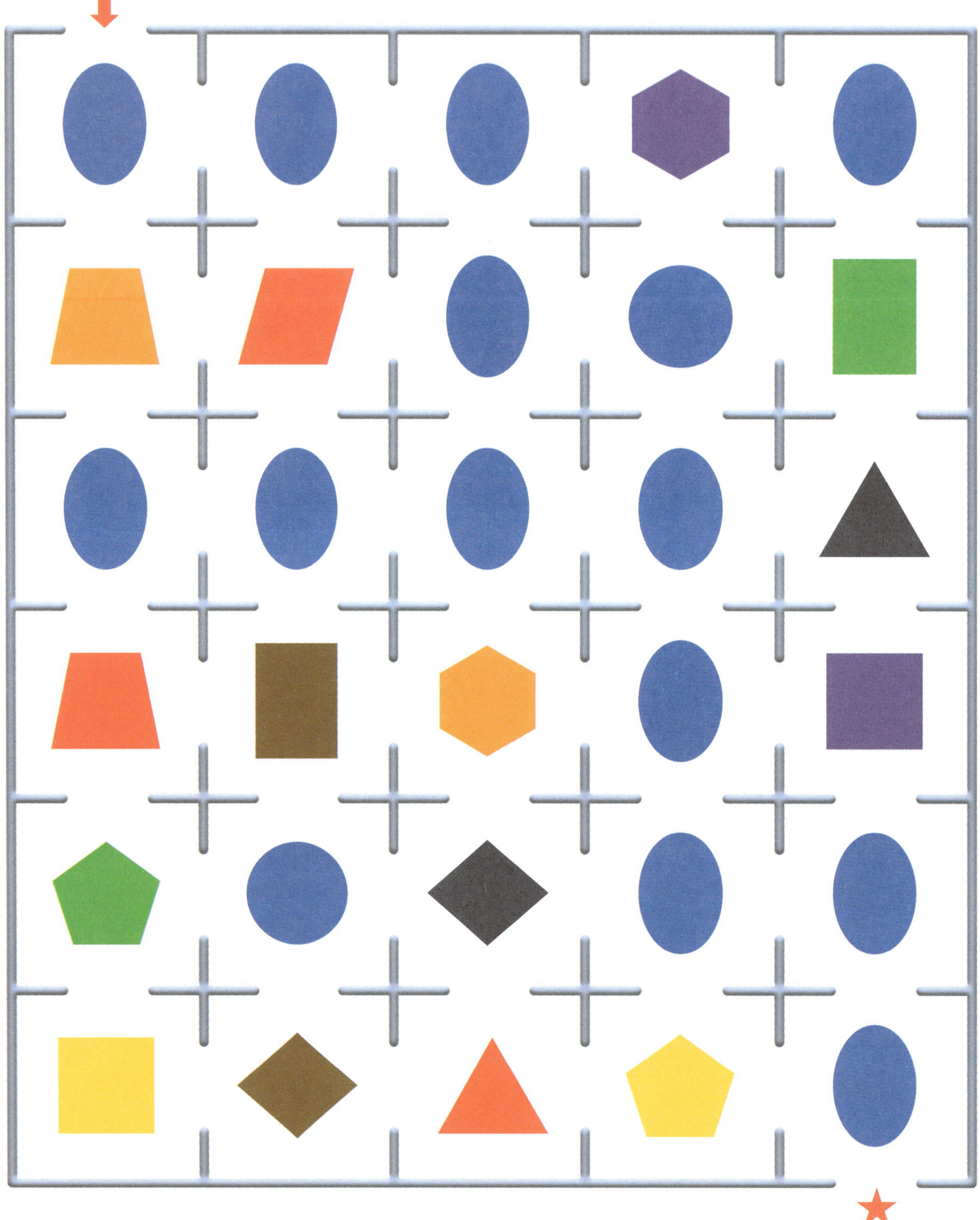

13 Giving Gifts

Name

Date

To parents
From this exercise on, each illustration includes multiple types of shapes. It is good to say the name of each shape and each object, as well as the colors aloud with your child.

■ Draw a line from one crayon to the matching crayon.

Square

Rectangle

To parents
From this exercise on, your child will follow the path while connecting multiple shapes in a pattern. It is important to have your child say the names of the shapes aloud as he or she connects them. If your child doesn't know what to do, please help him or her with the maze activity.

■ Draw a line from the arrow (⬇) to the star (★) by connecting the yellow squares, triangles, and circles (■▲●).

Name

Date

To parents
If your child can correctly draw the triangles same as each other, offer him or her a lot of praise for that. For example, you can say "The triangles you drew made the crown more beautiful!"

■ Draw a line from one crayon to the matching crayon.

Triangle Triangle Triangle

Diamond (Rhombus)

■ Draw a line from the arrow (↓) to the star (★) by connecting the orange rectangles, diamonds(rhombuses), and ovals (■◆⬮).

15 Balloon Travel

Name

Date

To parents
When your child has finished this exercise, talk with him or her about the objects they drew. You can also talk about more circle or oval shaped objects in your everyday life like lollipops, footballs, and so on.

- Draw a line from one crayon to the matching crayon.

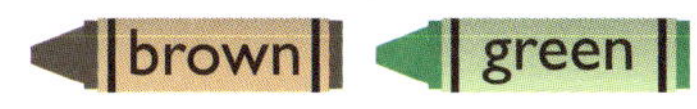

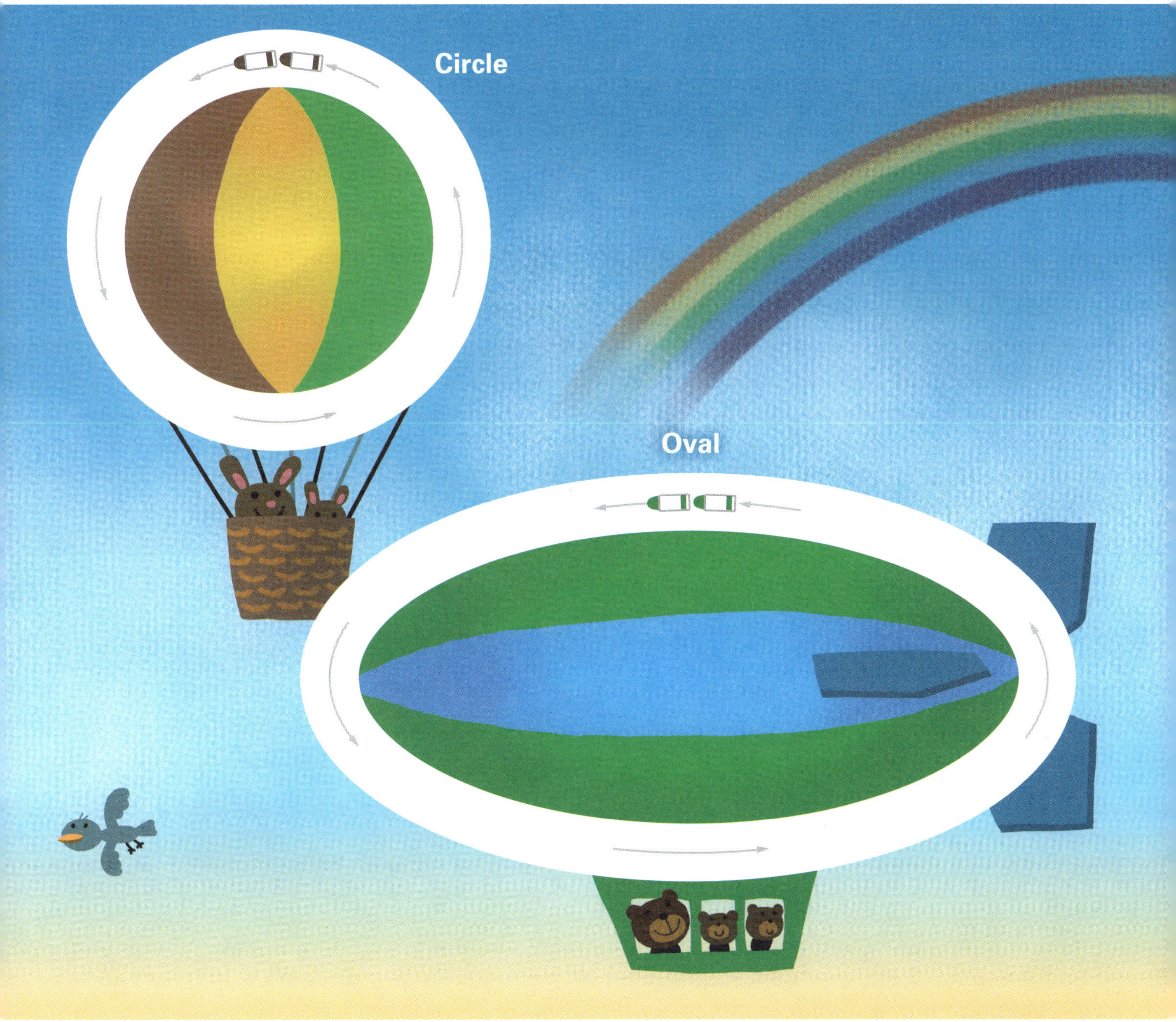

Circle

Oval

■ Draw a line from the arrow (⬇) to the star (★) by connecting the red triangles, ovals, and circles (▲⬮●).

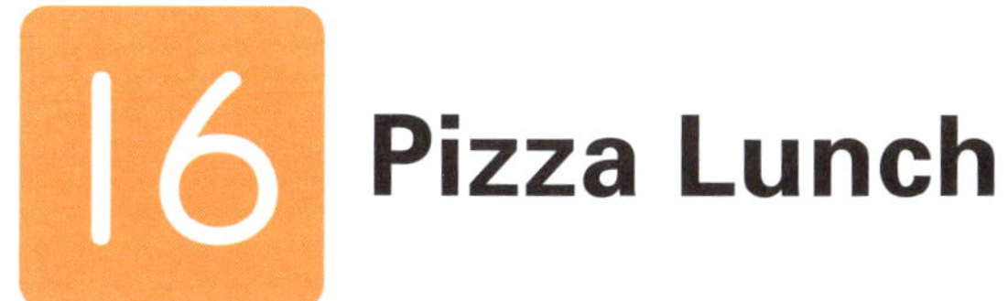

16 Pizza Lunch

Name
Date

To parents
In this exercise, your child will draw three different shapes to complete each object. When your child can distinguish between them, offer him or her a lot of praise for the hard work and the accomplishment.

Sample Response

■ Draw a line from one crayon to the matching crayon.

red brown blue

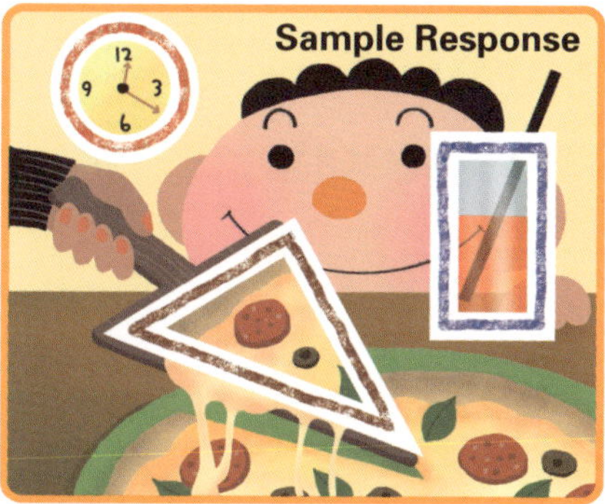

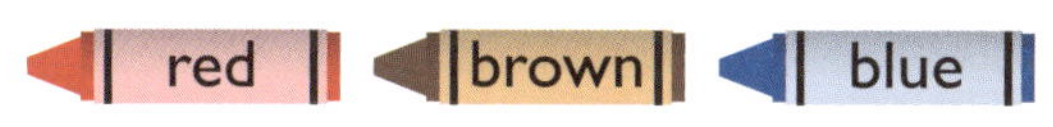

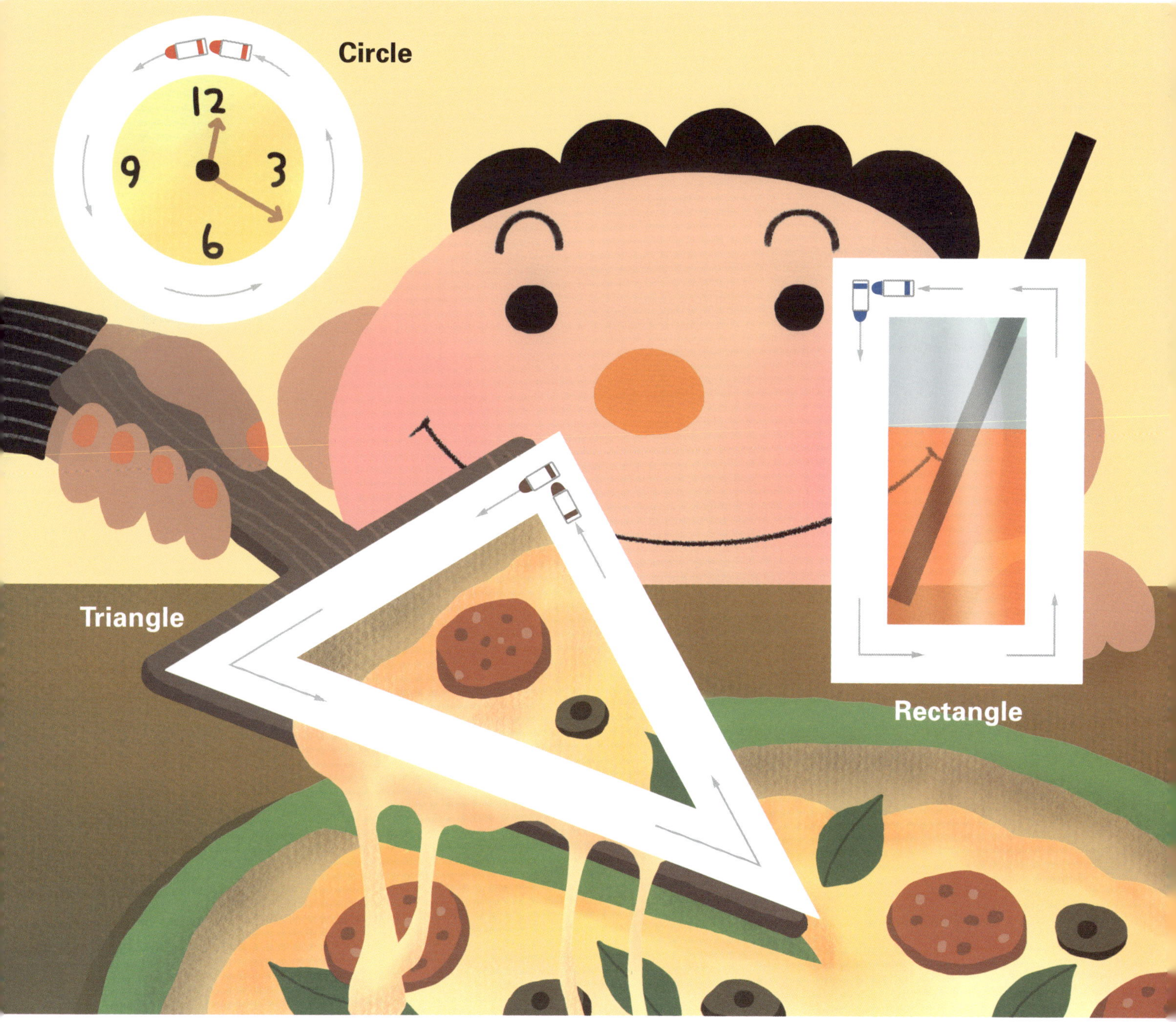
Circle
12
9
3
6
Triangle
Rectangle

■ Draw a line from the arrow (↓) to the star (★) by connecting the green diamonds (rhombuses), rectangles, and squares (◆ ▮ ■).

17 Parallelogram

Name

Date

To parents
From this exercise on, the shapes become more complicated. However, your child will gradually and smoothly be able to recognize them through the activities.

■ Draw a line from the dot (●) to the star (★).

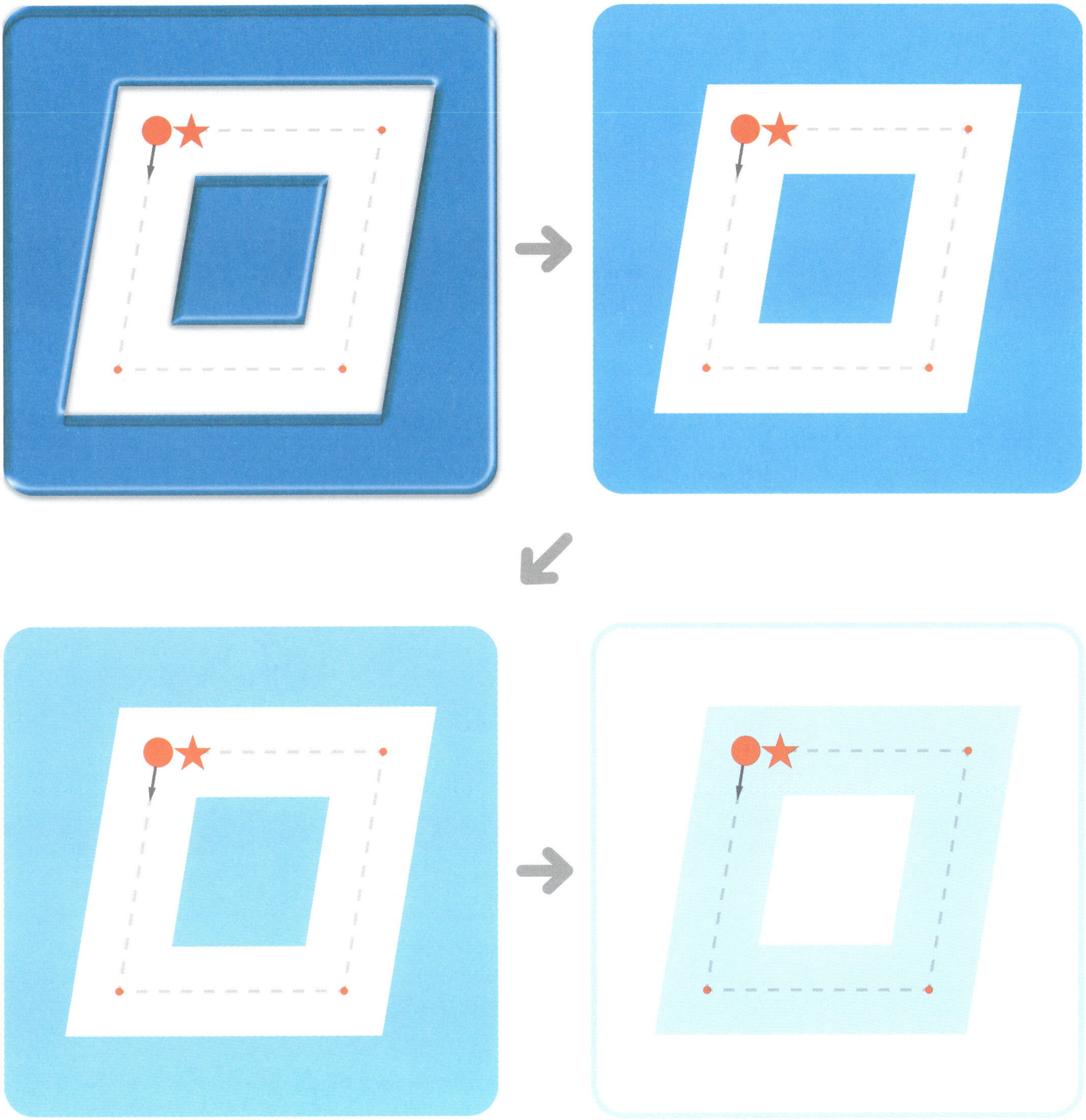

■ Draw a line from the dot (●) to the star (★).

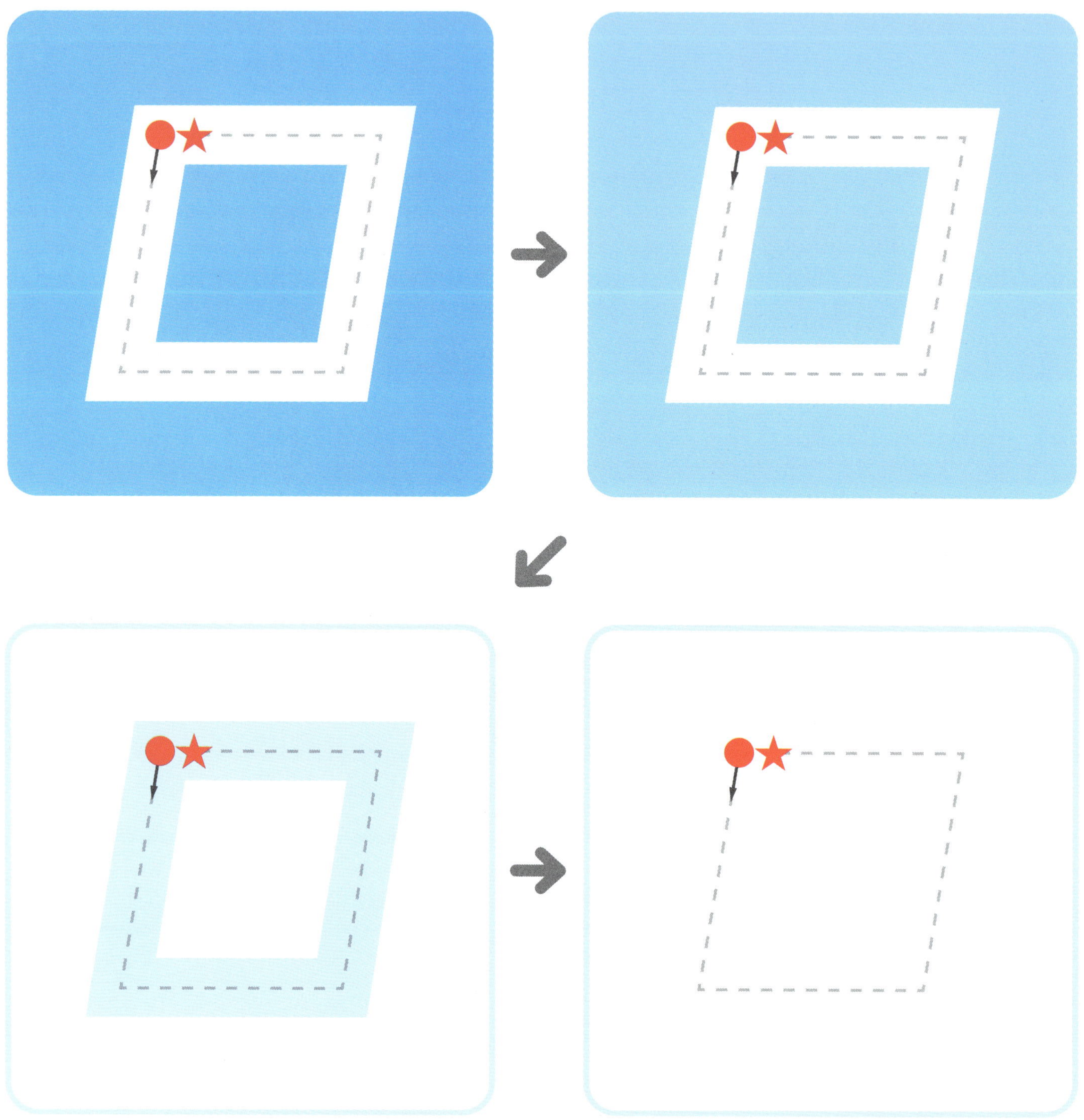

18 Parallelograms: Garden Boxes

Name

Date

To parents

When your child has finished this exercise, talk with him or her about the objects and the colors. For example, you can say "Have you seen them outside before? Can you tell me what colors they are?"

■ Draw a line from one crayon to the matching crayon.

Parallelogram

Parallelogram

■ Draw a line from the arrow (↓) to the star (★) by connecting the violet (purple) parallelograms (▰).

Trapezoid

Name

Date

To parents
It is okay if your child cannot draw the shape in one stroke at first. The important thing is that he or she can learn the characteristics of the shape. When your child has finished the exercise, offer him or her a lot of praise.

■ Draw a line from the dot (●) to the star (★).

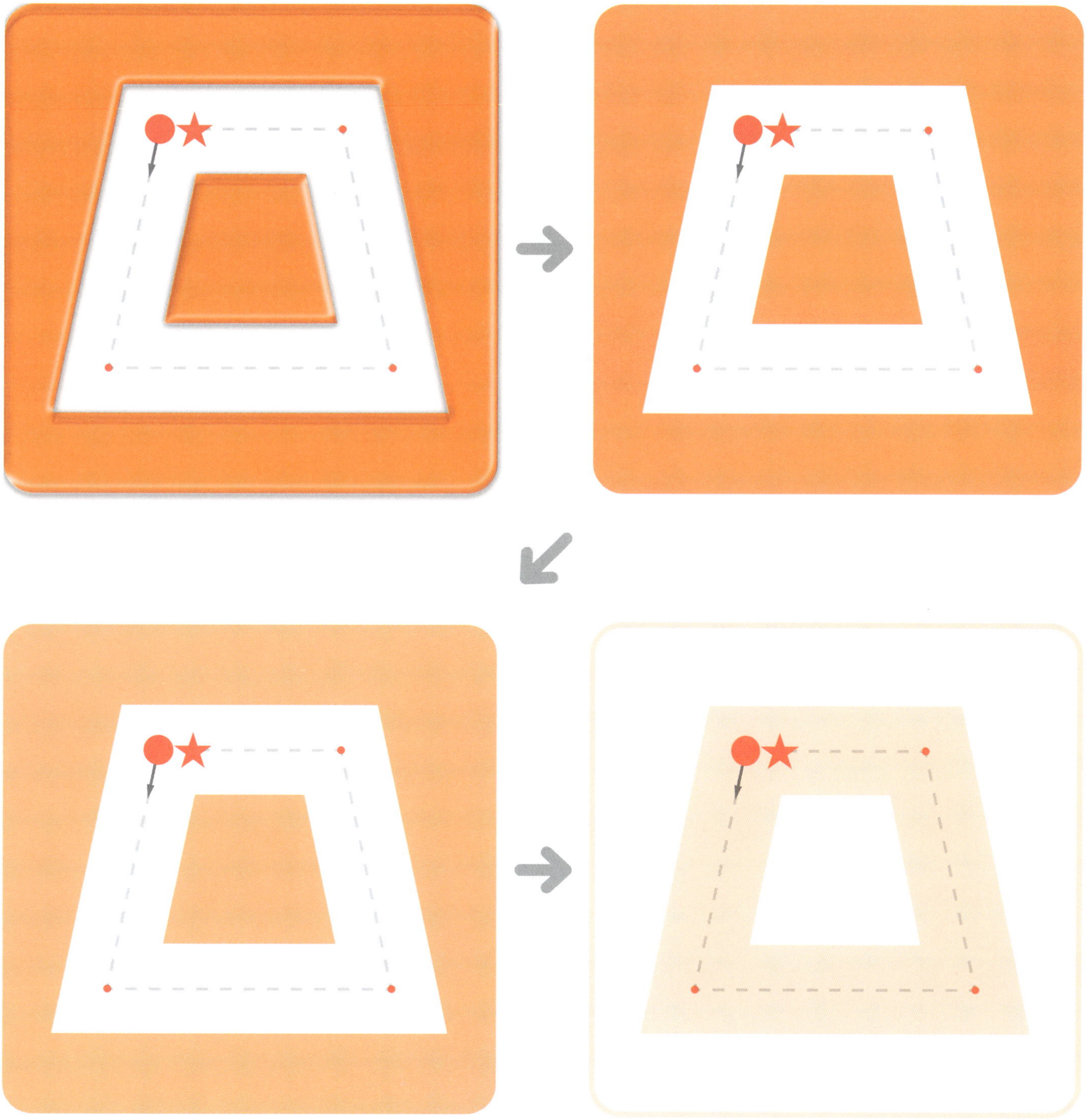

■ Draw a line from the dot (●) to the star (★).

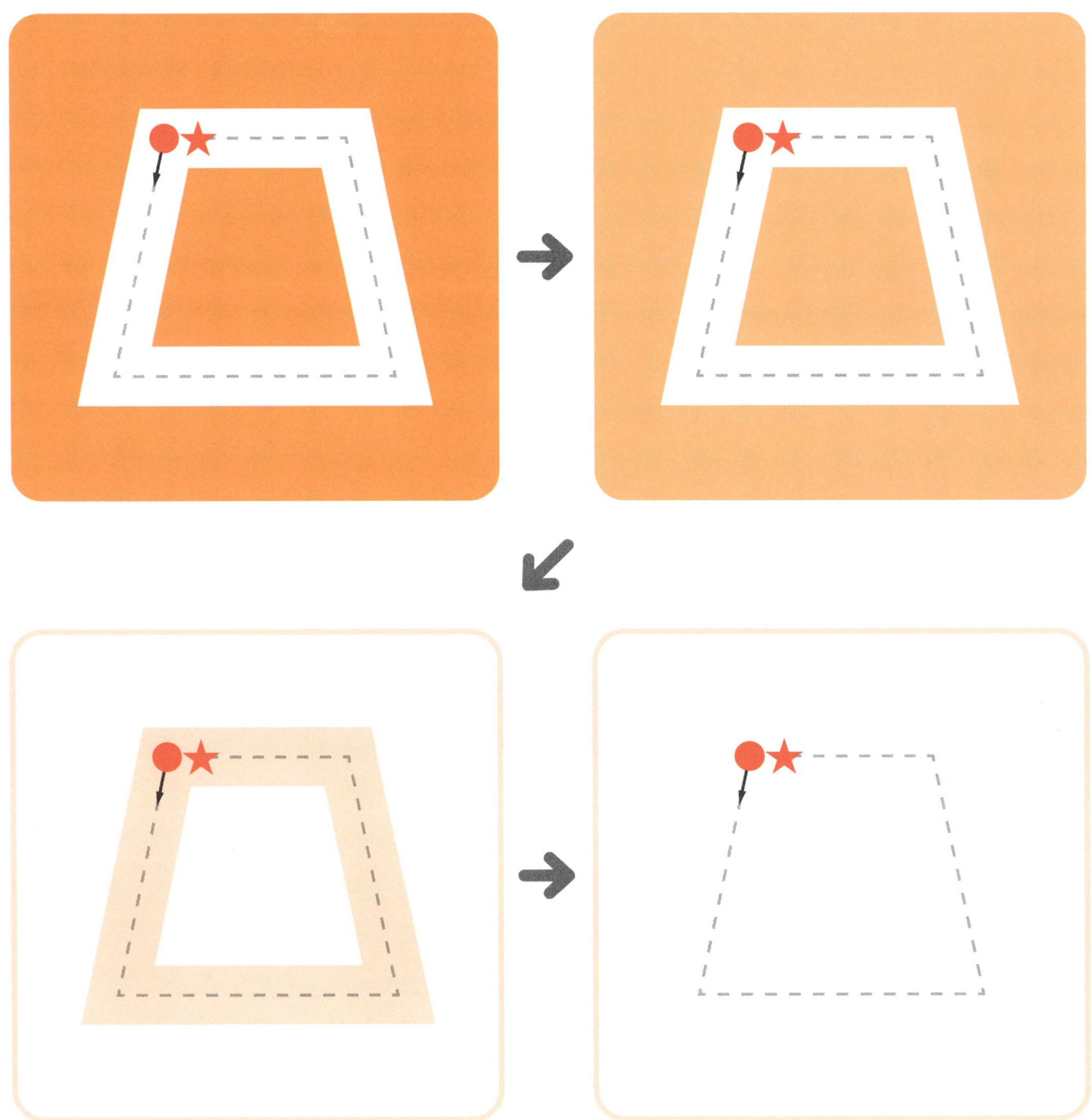

20 Trapezoids: Grocery Store

Name

Date

To parents
After your child has finished the exercise, it is good to ask him or her what the shape's name and what the color is. You can also ask when he or she could see these objects in everyday life.

■ Draw a line from one crayon to the matching crayon.

violet blue

Trapezoid

3

Trapezoid

Draw a line from the arrow (↓) to the star (★) by connecting the red trapezoids (▰).

21 Pentagon

Name

Date

To parents
If your child is having difficulty drawing the shape, please remind him or her to draw slowly and carefully. It might be good to ask your child, what does this shape look like?

■ Draw a line from the dot (●) to the star (★).

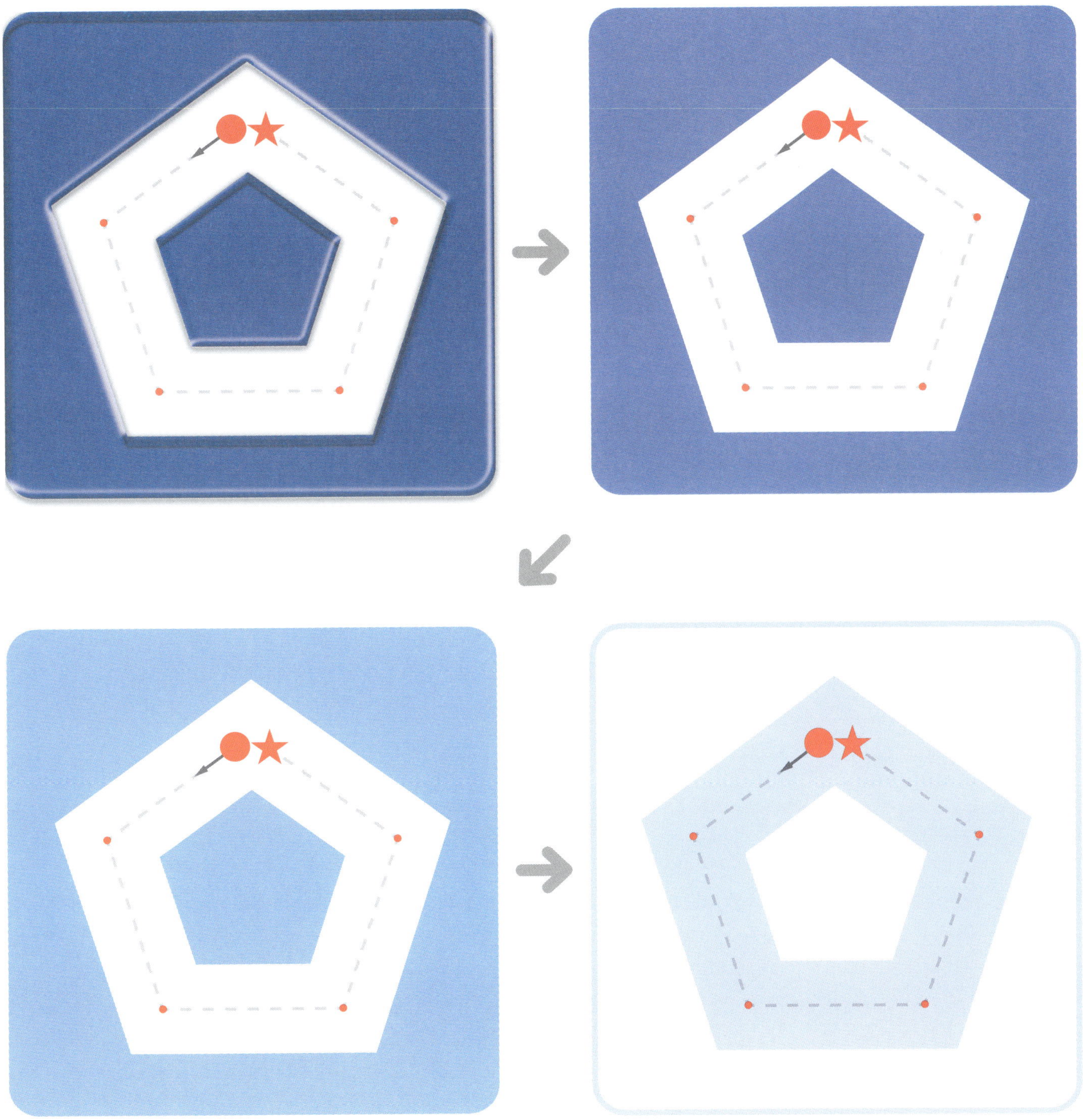

■ Draw a line from the dot (●) to the star (★).

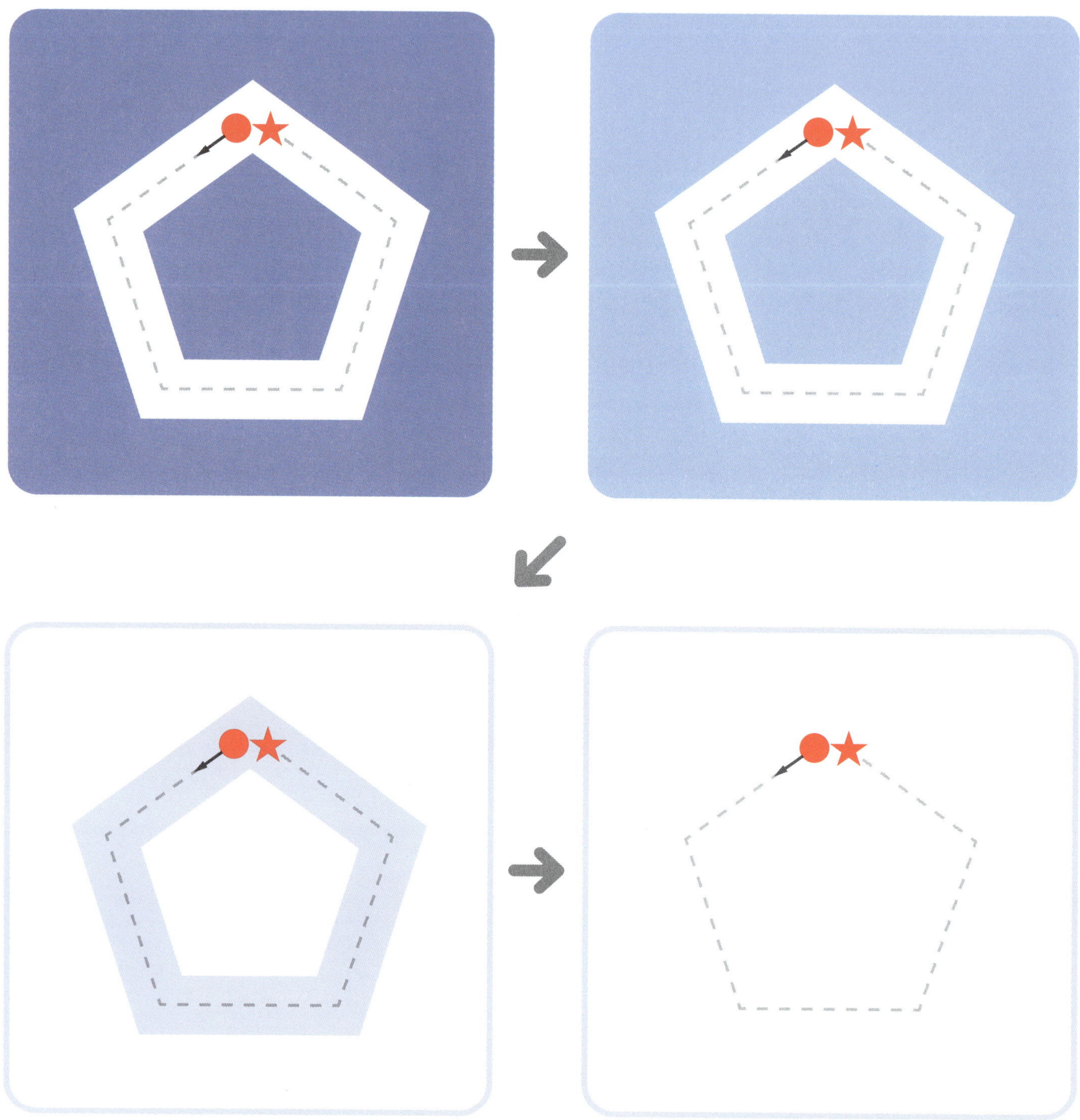

22 Pentagons: Bird Houses

Name

Date

To parents
When your child has finished this exercise, talk with him or her about the objects and the colors. For example, you can say "Which bird house do you like, the small yellow one or the big red one?"

■ Draw a line from one crayon to the matching crayon.

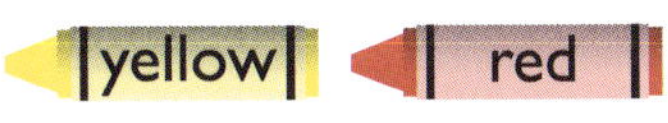

Pentagon

Pentagon

■ Draw a line from the arrow (↓) to the star (★) by connecting the yellow pentagons (⬟).

23 Hexagon

Name

Date

To parents
This is the last shape your child will learn individually in this workbook. It is okay if your child cannot draw the shape in one stroke at first. The important thing is that he or she can learn the characteristics of the shape.

■ Draw a line from the dot (●) to the star (★).

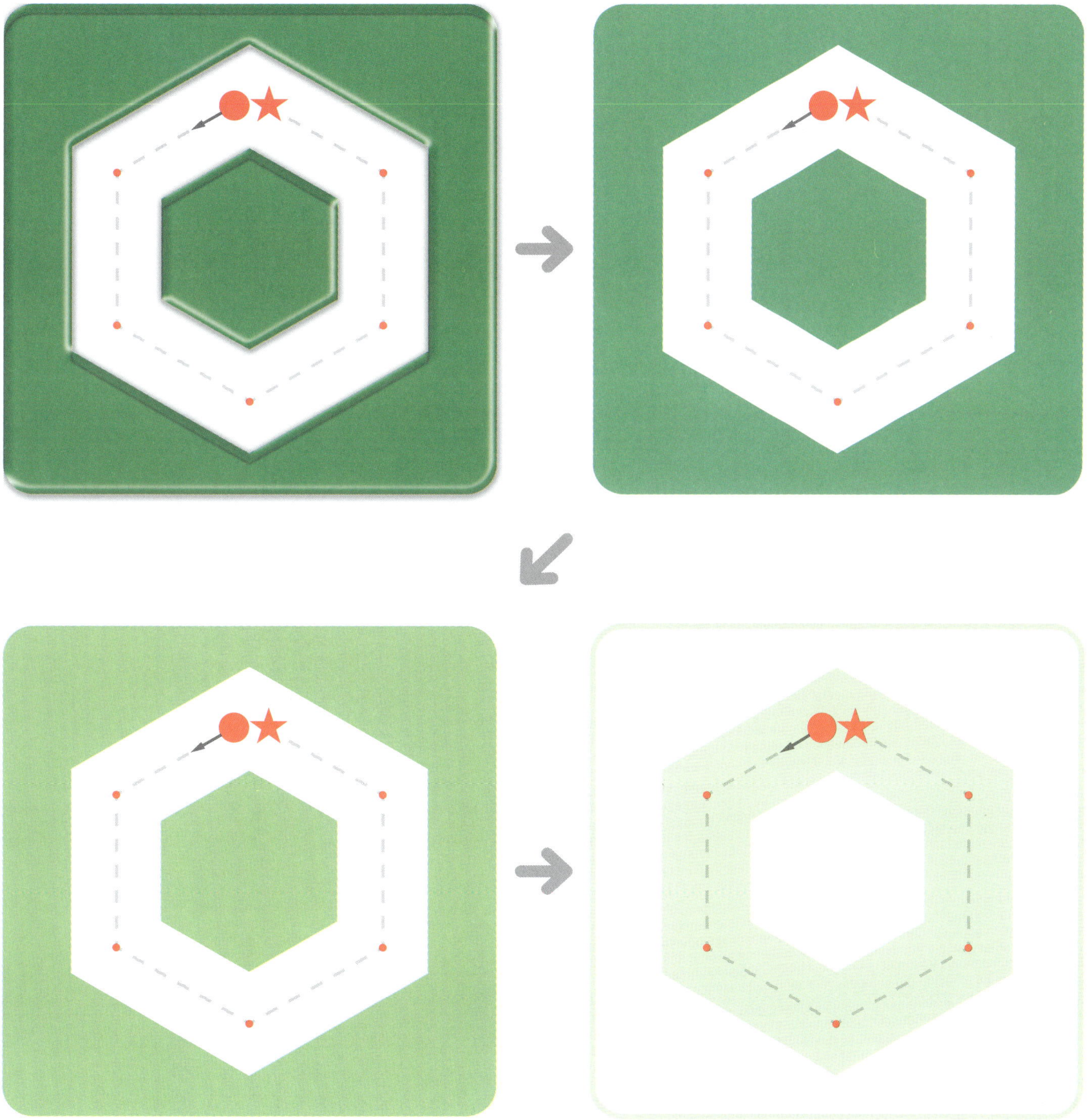

Draw a line from the dot (●) to the star (★).

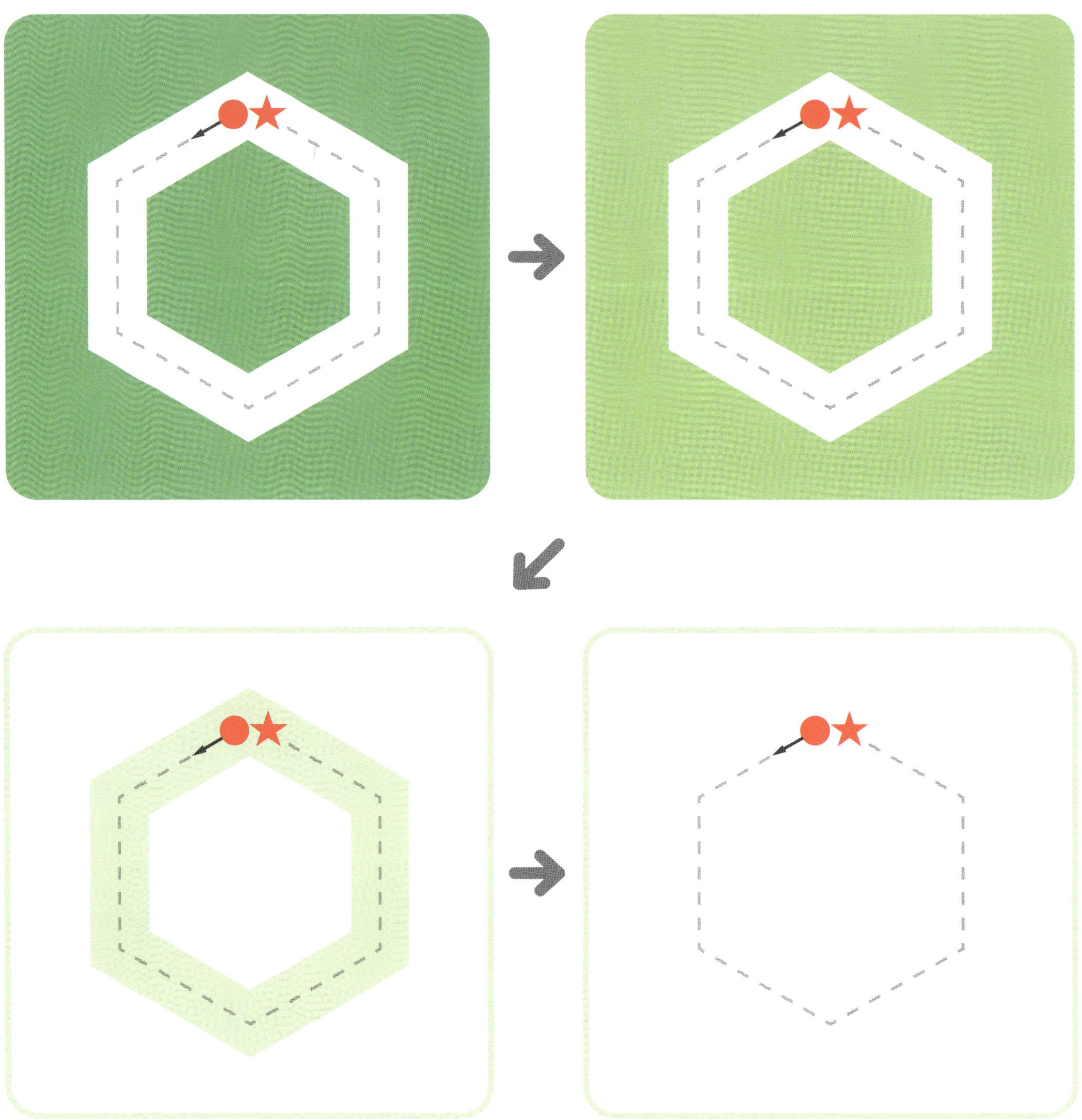

24 Hexagons: Snowflakes

Name

Date

To parents
When your child has finished this exercise, talk with him or her about the objects and the colors. For example, you can say "Have you seen them outside before?"

■ Draw a line from one crayon to the matching crayon.

Hexagon

Hexagon

■ Draw a line from the arrow (↓) to the star (★) by connecting the black hexagons (⬢).

25 Sea Travel

Name

Date

To parents

From this exercise on, your child will learn an object consists of the multiple shapes. If your child can see that the green square and the orange trapezoid make the ship, offer him or her a lot of praise.

■ Draw a line from one crayon to the matching crayon.

green orange yellow

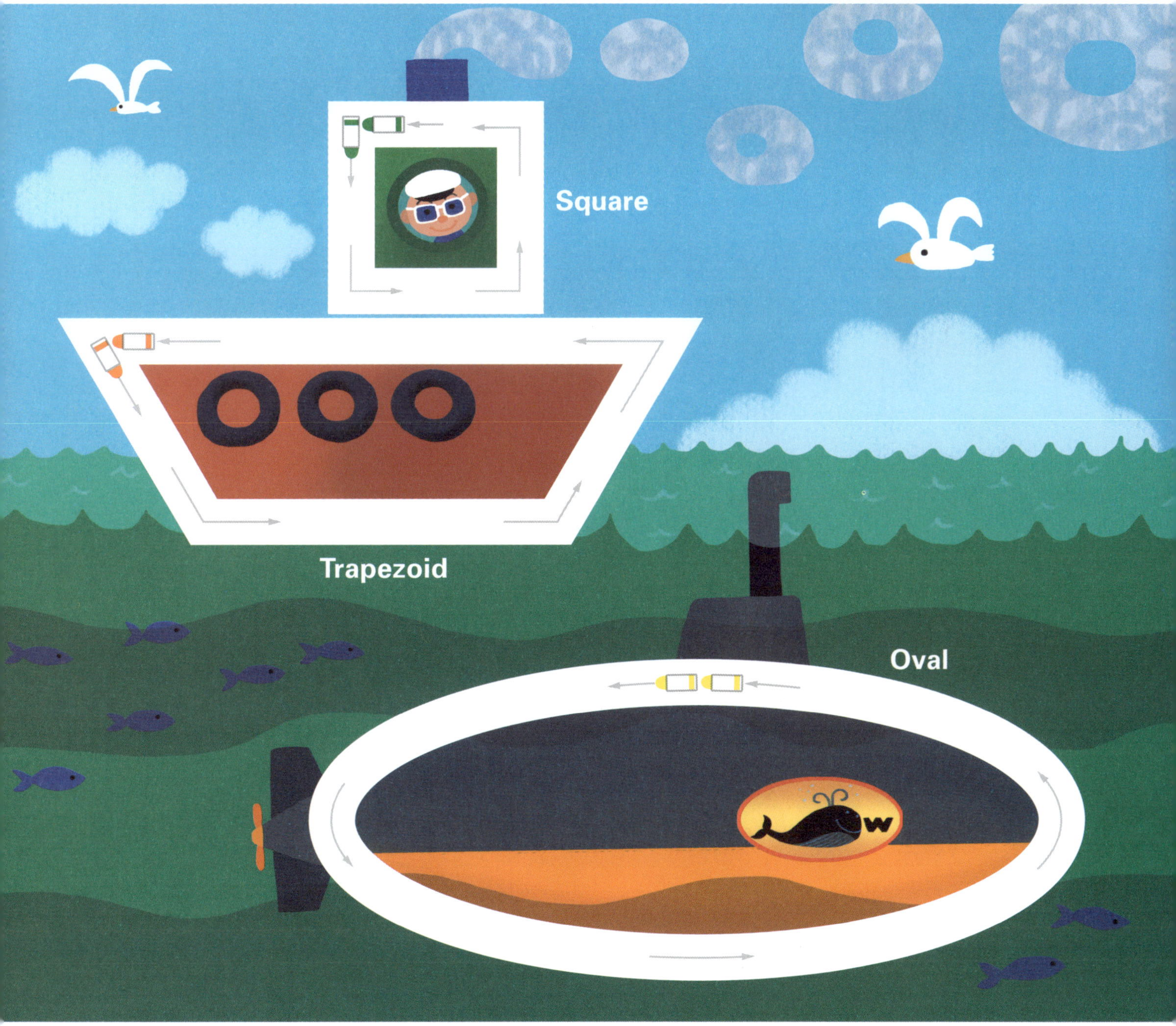

Square

Trapezoid

Oval

Draw a line from the arrow (↓) to the star (★) by connecting the blue parallelograms and trapezoids (▰ ▲).

26 Jazz Singer

Name

Date

To parents
The illustration shows that a microphone can be made with a circle and a rectangle. Encourage your child to think of other objects that consist of multiple shapes in his or her everyday life.

Sample Response

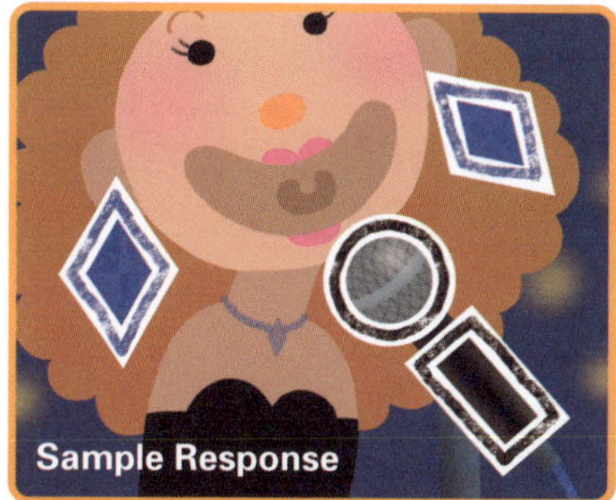

■ Draw a line from one crayon to the matching crayon.

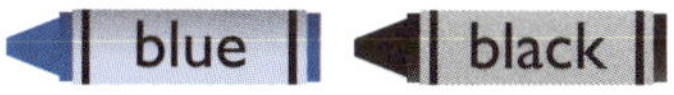
blue
black

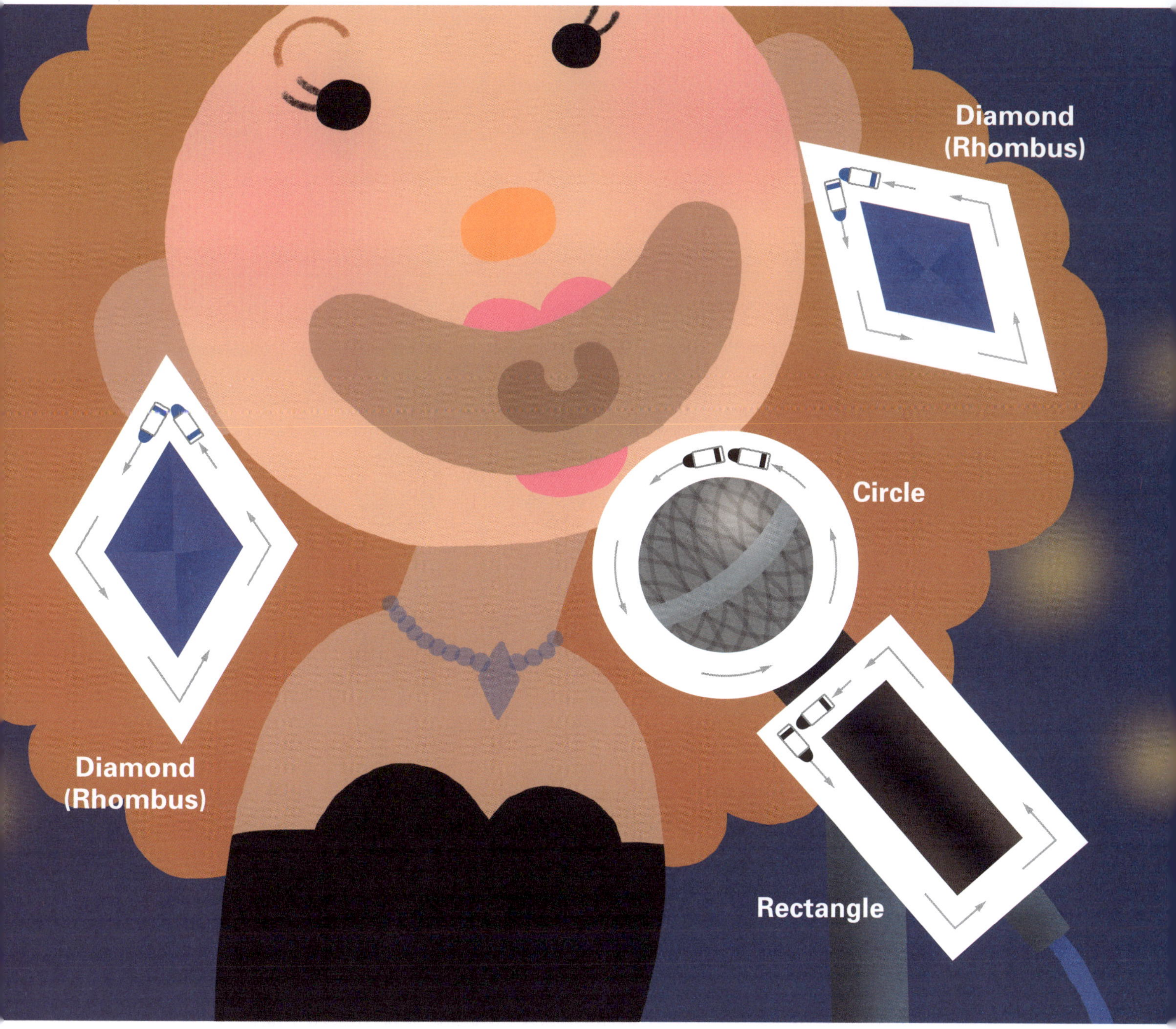
Diamond (Rhombus)
Circle
Diamond (Rhombus)
Rectangle

■ Draw a line from the arrow (⬇) to the star (★) by connecting the violet(purple) pentagons and hexagons (⬟ ⬢).

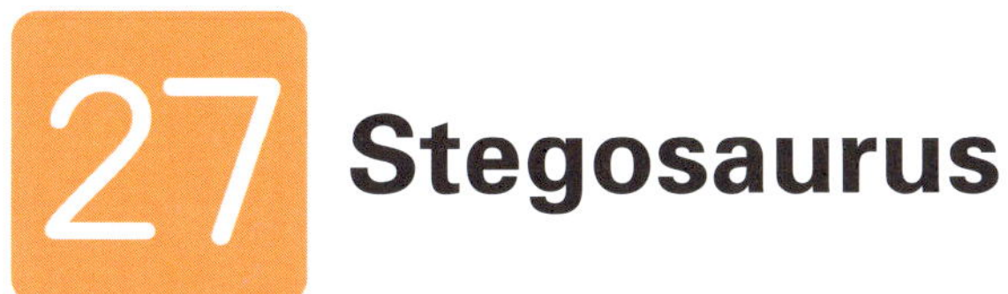

27 Stegosaurus

Name

Date

To parents
It is good to say the name of shapes and the objects, as well as the colors aloud with your child. For example, you can say "The stegosaurus has many red plates on its back and has a big green body."

Sample Response

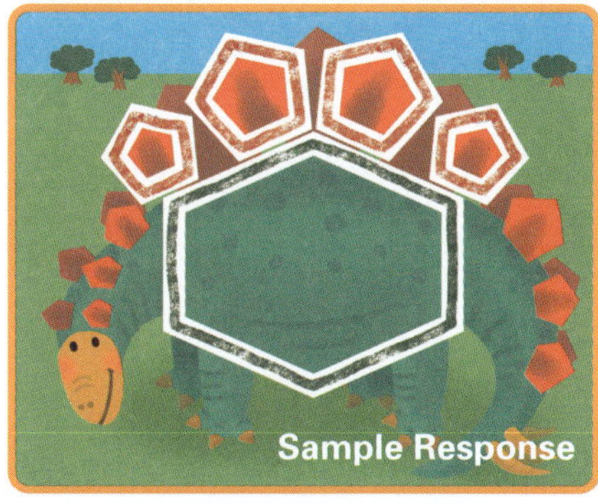

■ Draw a line from one crayon to the matching crayon.

red
green

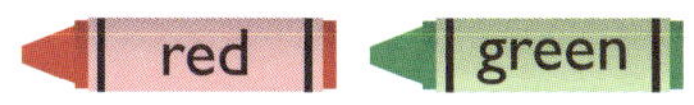

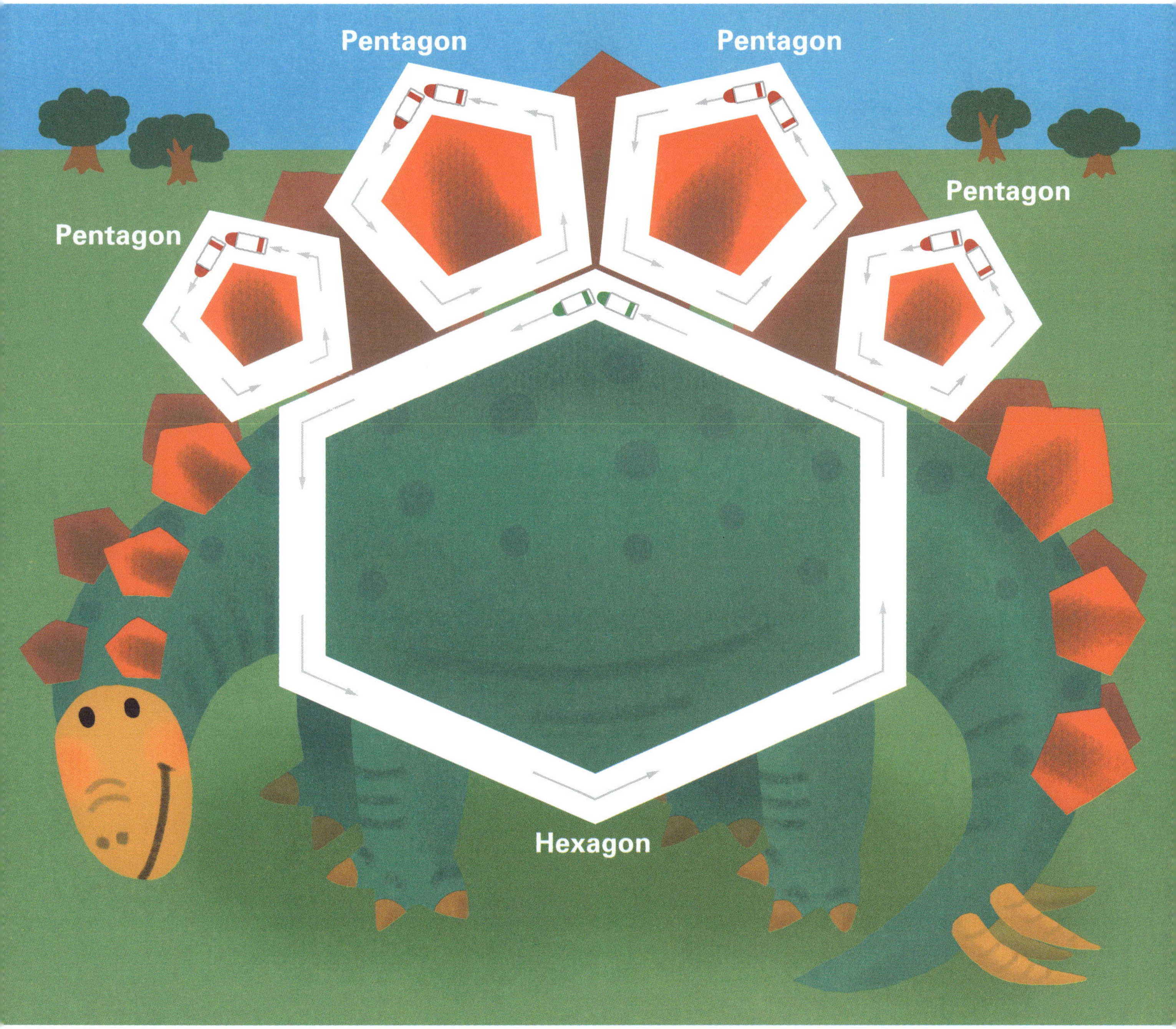
Pentagon
Pentagon
Pentagon
Pentagon
Hexagon

■ Draw a line from the arrow (↓) to the star (★) by connecting the brown parallelograms, trapezoids, pentagons, and hexagons (▰ ▲ ⬟ ⬢).

28 Rocket Ship

Name

Date

To parents
Your child will be reaching the latter part of this workbook soon. He or she has repeatedly learned shapes and colors so far. Encourage your child to keep working at his or her own pace.

Sample Response

■ Draw a line from one crayon to the matching crayon.

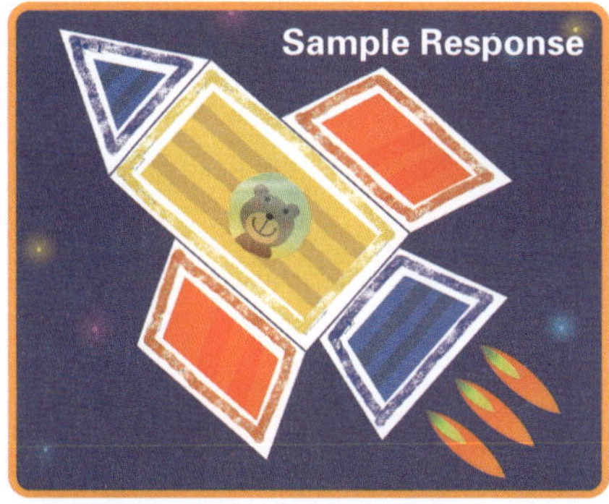

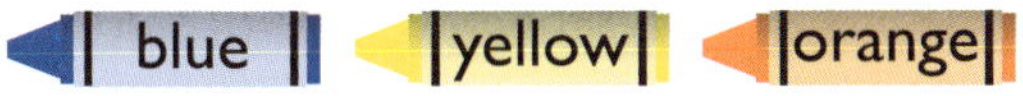

blue
yellow
orange

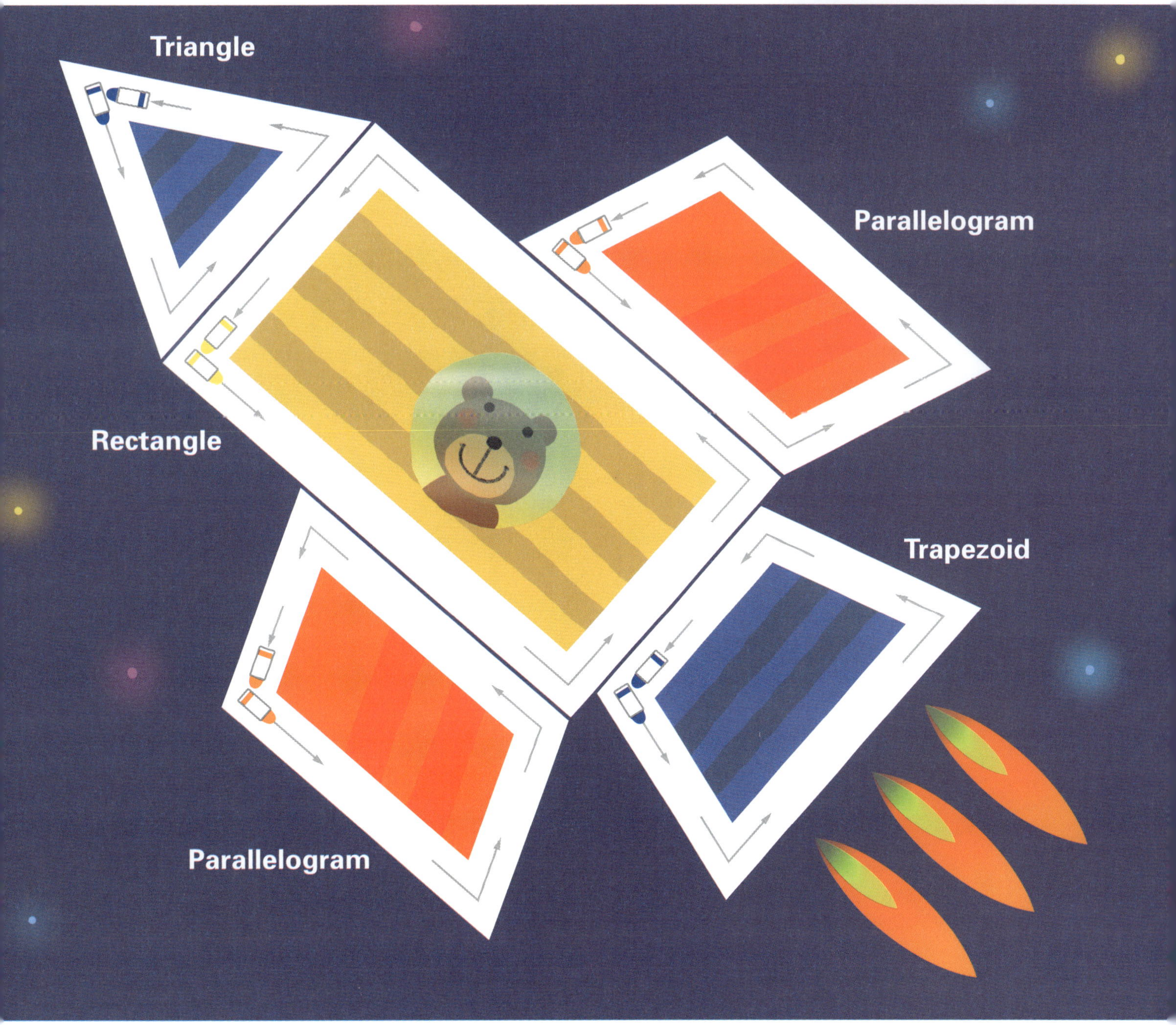

Triangle
Parallelogram
Rectangle
Trapezoid
Parallelogram

■ Draw a line from the arrow (↓) to the star (★) by connecting the black pentagons, trapezoids, hexagons, and parallelograms (⬟ ⏢ ⬢ ▰).

29 Review

Name

Date

To parents
The following two pages offer exercises to review the shapes and the colors previously learned. When your child can correctly draw the shapes on his or her own, offer a lot of praise for the accomplishment.

■ Draw a red square and a blue triangle.

red **square**

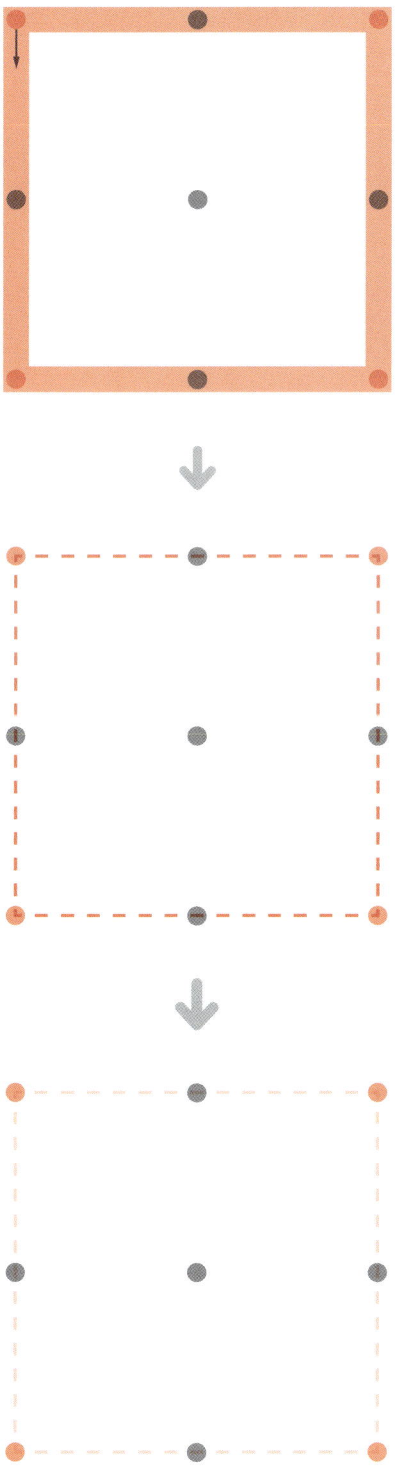

blue **triangle**

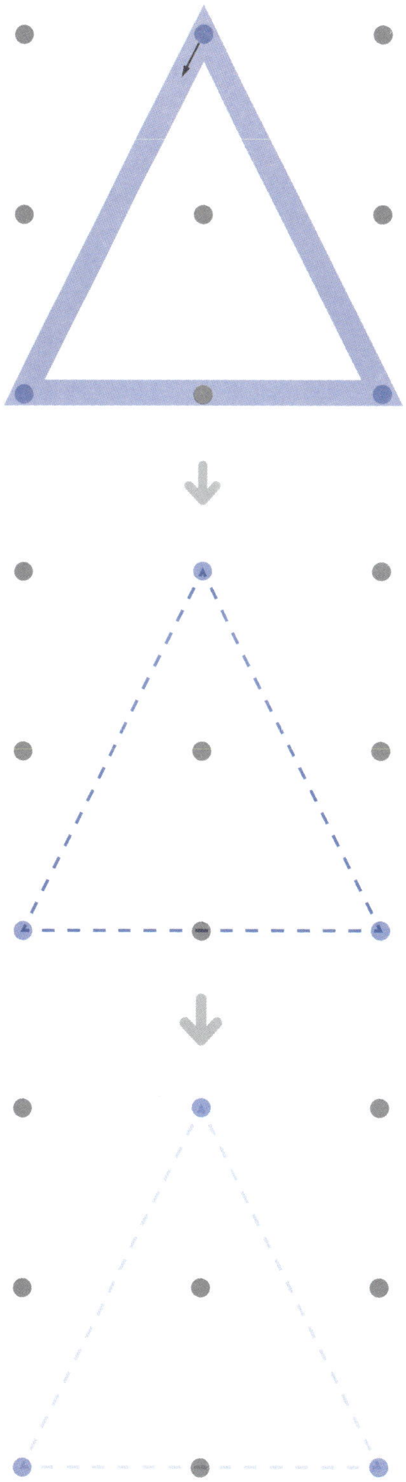

■ Draw a green rectangle and an orange diamond (rhombus).

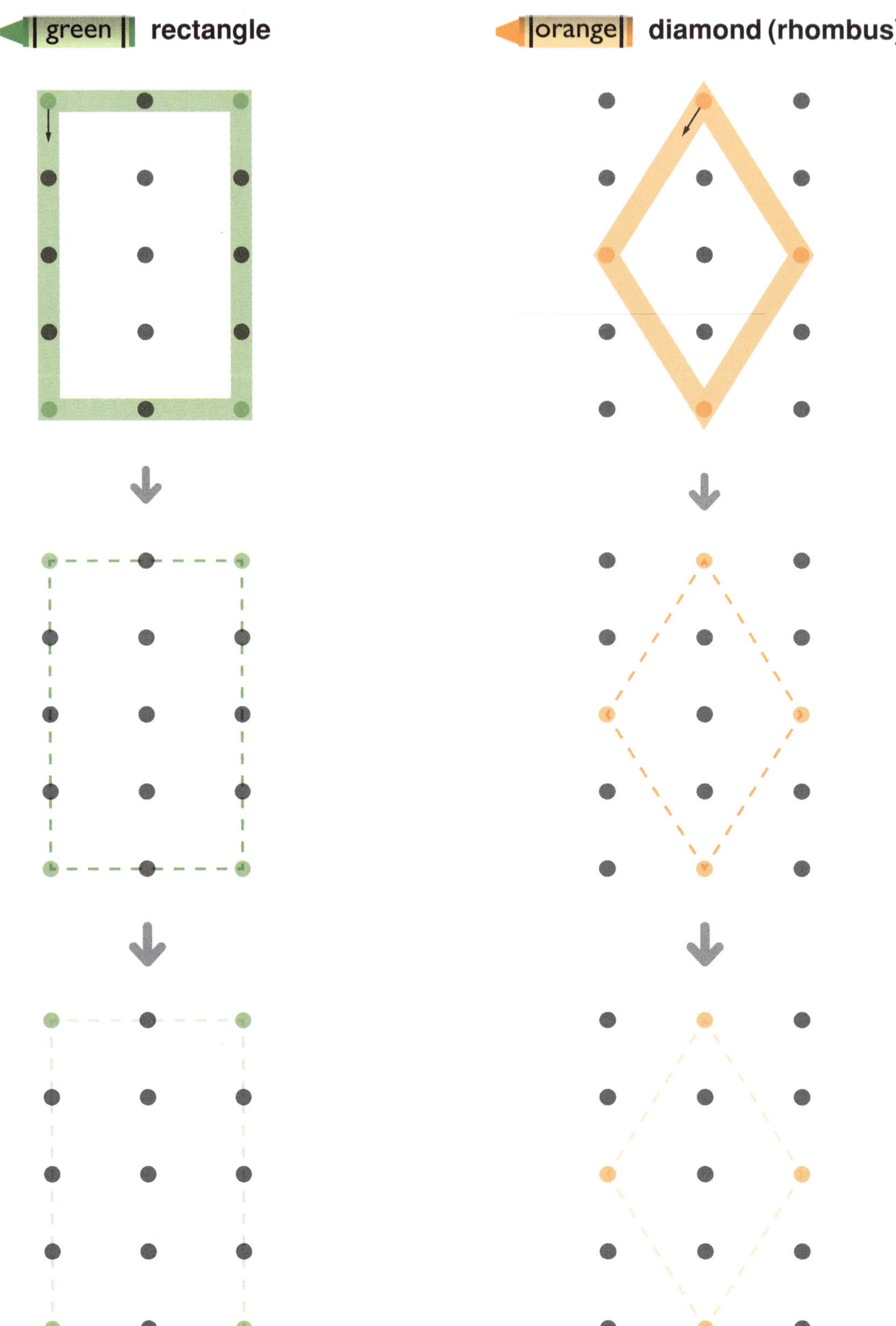

30 Review

Name

Date

To parents
It may be difficult for your child to draw a long line to connect the dots. If necessary, guide your child's hand and draw together to demonstrate.

■ Draw a violet (purple) parallelogram and a yellow trapezoid.

violet **parallelogram**

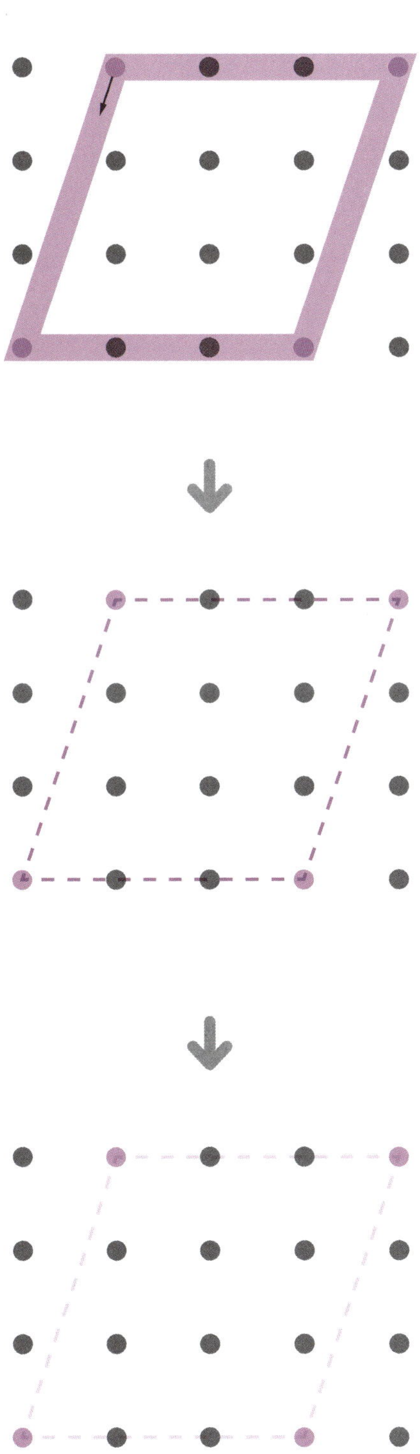

yellow **trapezoid**

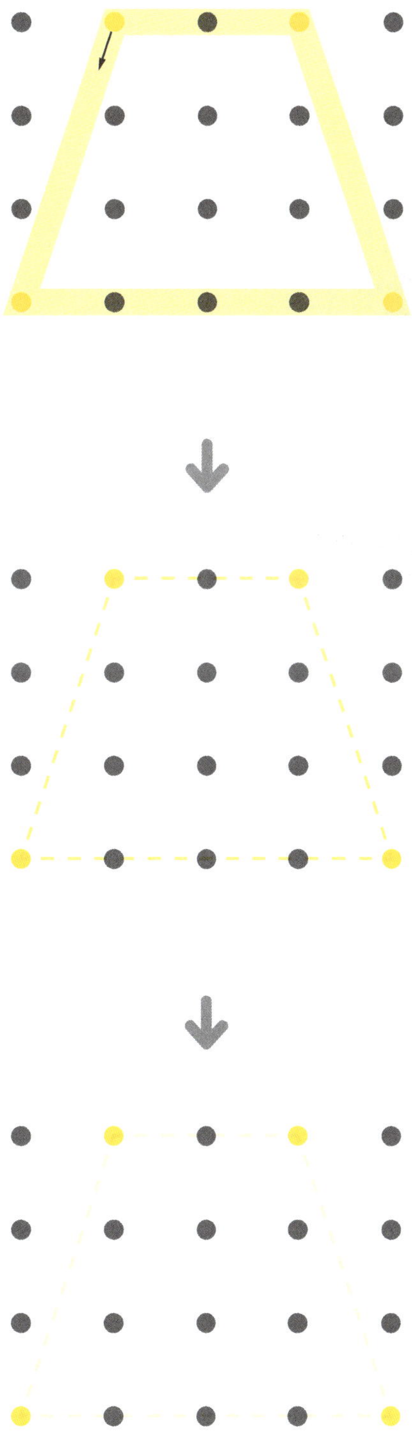

Draw a brown pentagon and a black hexagon.

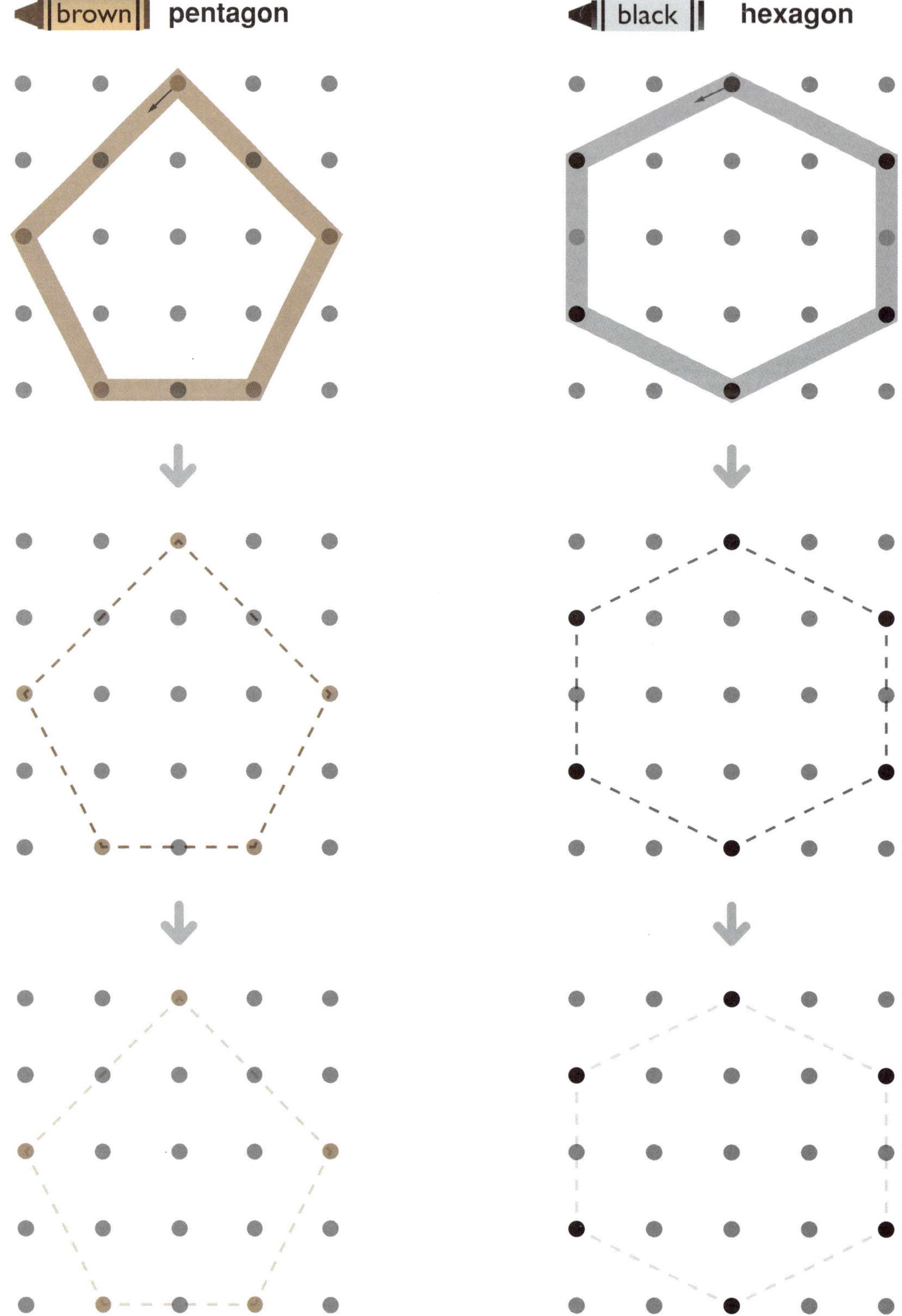

31 Which is the Same Shape and Color?

Name

Date

To parents
This exercise is to help your child recognize different sizes and orientations of the same shape using designated colors. From now until the end of the workbook, there are answers provided for each exercise on the last two pages of this book (pp. 77-78).

■ Circle the shapes below that are the same shape and color as the sample.

Sample

red
circle

What Is It?

To parents
The following backside pages are color-by-letter activities. If necessary, please point out the sections to be colored. Offer your child a lot of praise when he or she finishes each activity.

■ Use the key below to color by letter.
b=brown y=yellow

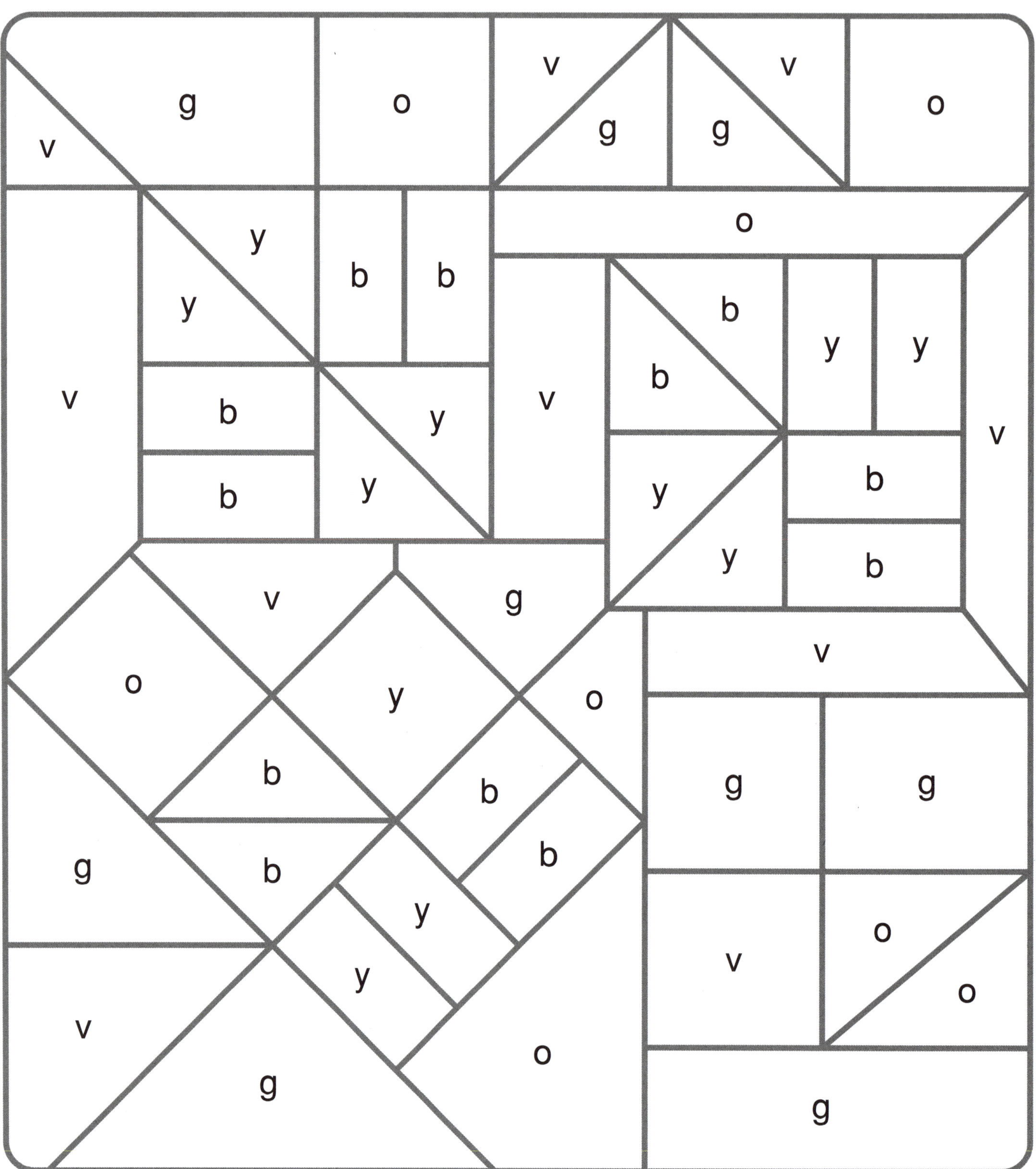

32 Which is the Same Shape and Color?

Name

Date

To parents
If your child circles the incorrect color shapes, please advise him or her to check the samples carefully.

■ Circle the shapes below that are the same shape and color as the sample.

What Is It?

To parents
In this activity, the sections to be colored are the shapes your child has learned in this workbook. Offer your child a lot of praise if he or she notices this or recognizes a specific shape.

■ Use the key below to color by letter.
o=orange v=violet (purple)

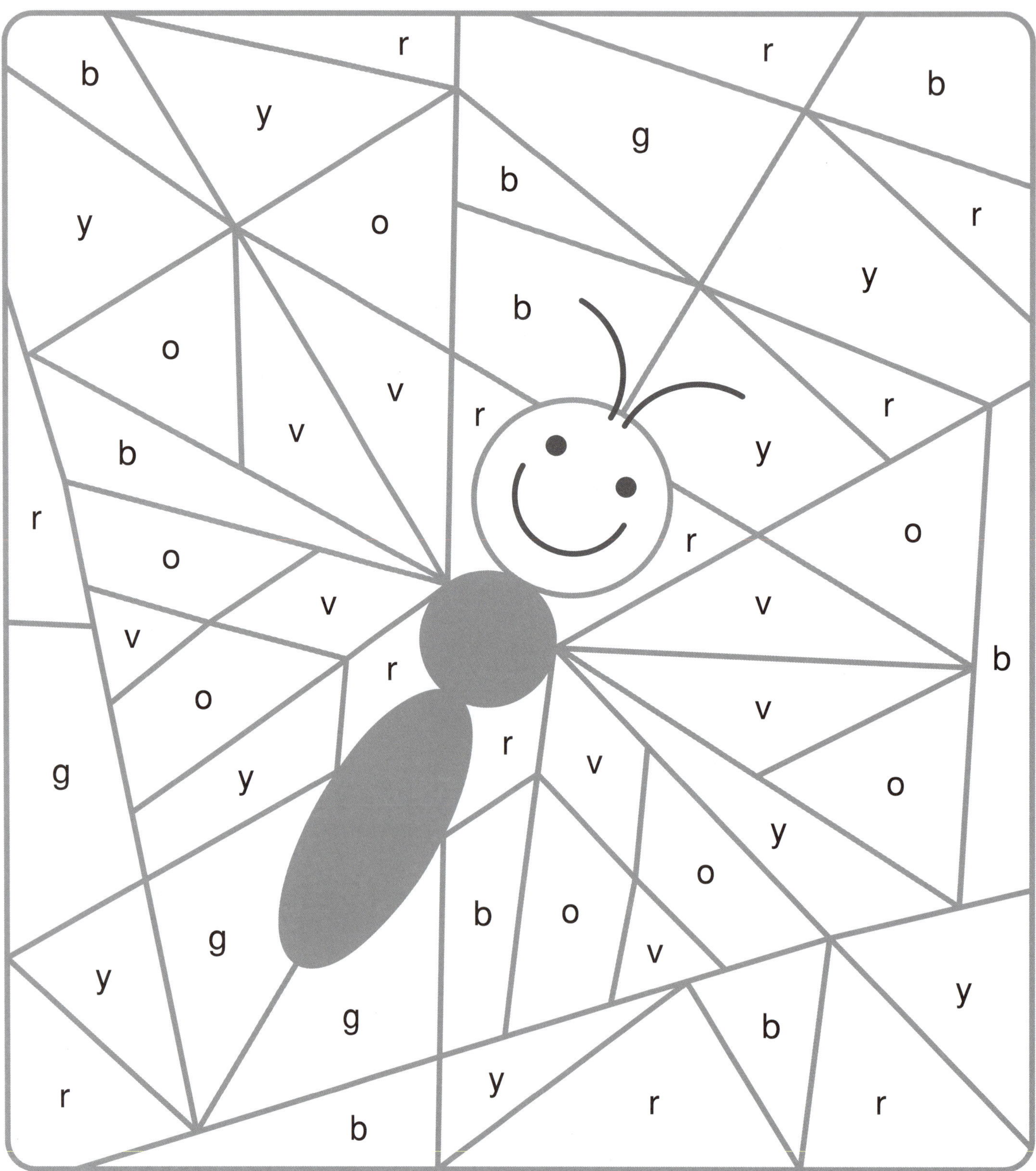

33 Which is the Same Shape and Color?

Name

Date

To parents
If your child has difficulty finding the correct shape, please remind him or her of the characteristics of the shape and the correct color.

■ Circle the shapes below that are the same shape and color as the sample.

Sample

green oval

black parallelogram

orange rectangle

What Is It?

■ Use the key below to color by letter.
r=red b=blue o=orange g=green

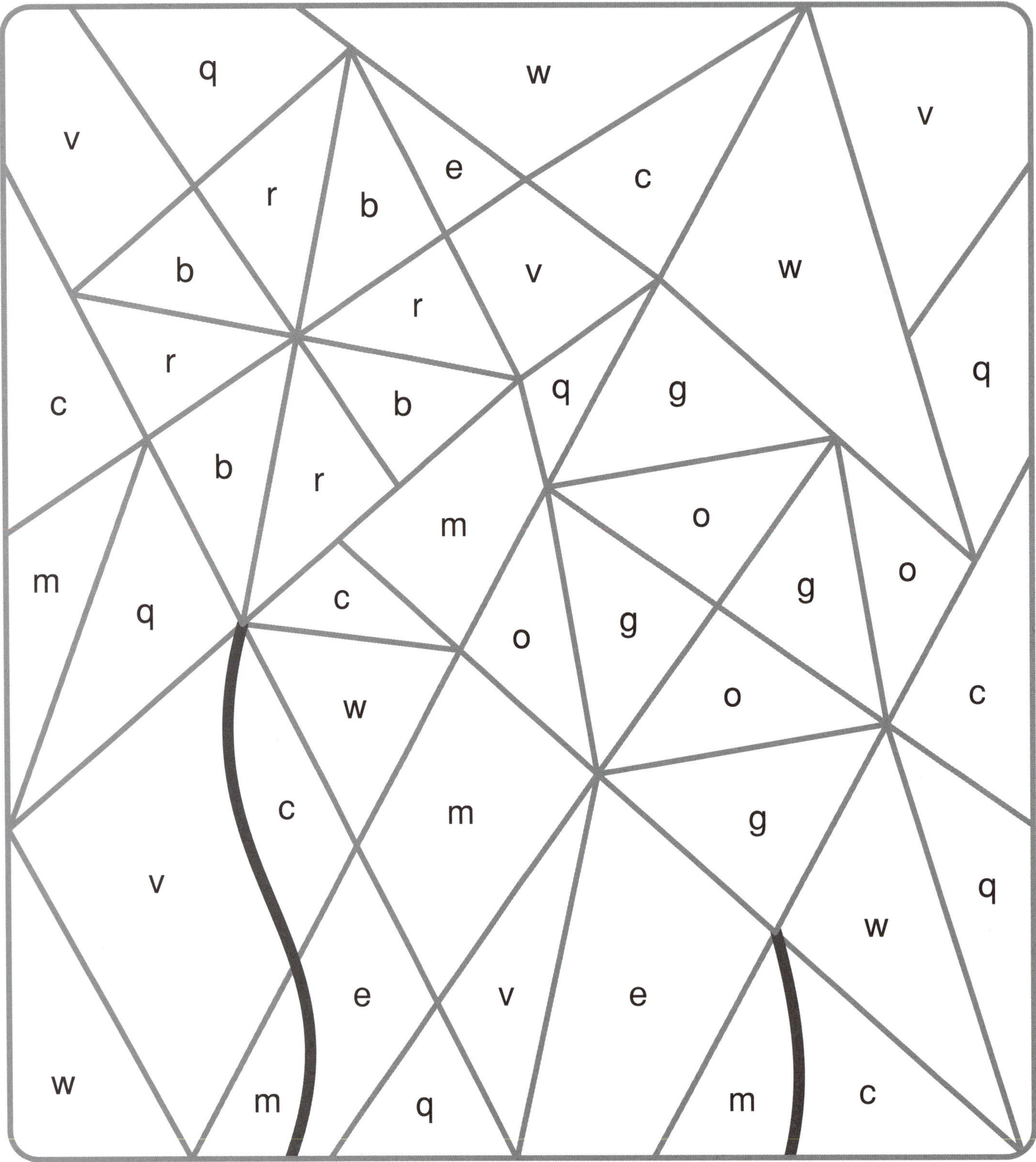

34 Which is the Same Shape and Color?

Name

Date

To parents
If your child circles the incorrect color shapes, please advise him or her to check the samples carefully.

■ Circle the shapes below that are the same shape and color as the sample.

Sample

brown trapezoid

violet (purple) hexagon

red diamond (rhombus)

blue triangle

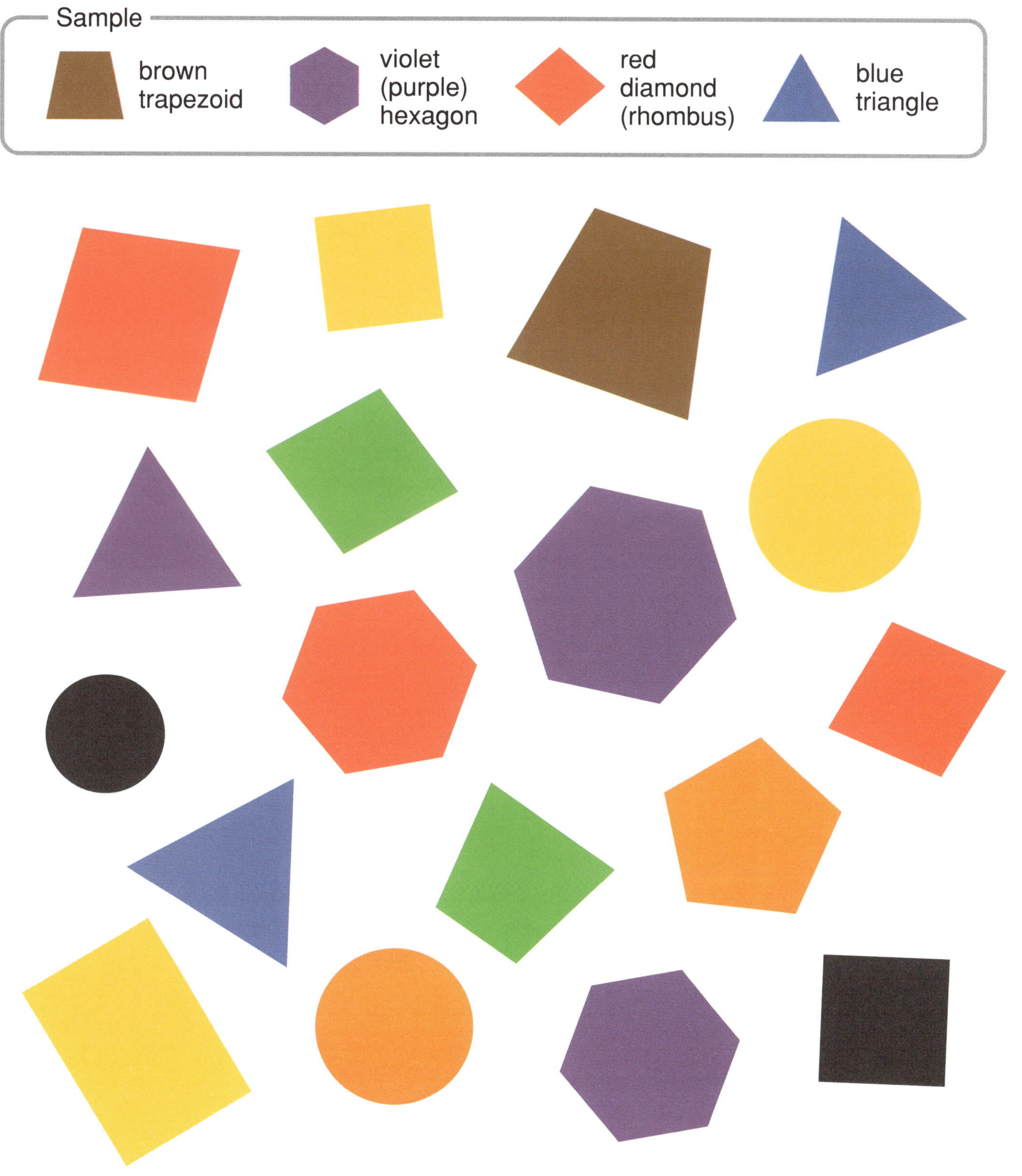

What Is It?

- Use the key below to color by letter.
  r=red b=brown y=yellow

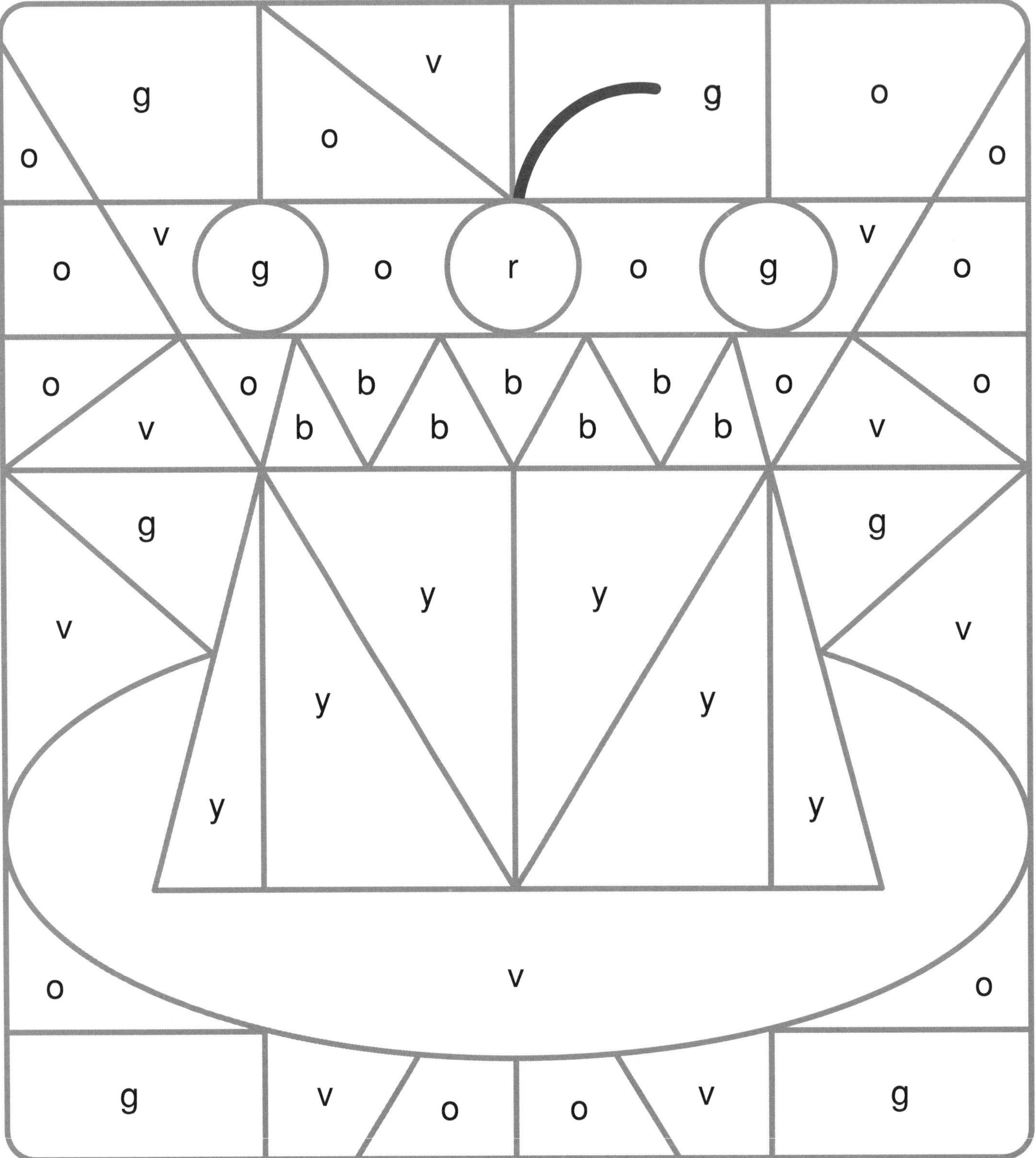

35 Which is the Same Shape and Color?

Name

Date

To parents
Like the previous pages, this exercise is also designed to help your child recognize shapes that are the same but are different sizes and orientations. If your child has difficulty finding the correct shape, please remind him or her of the characteristics of the shape.

■ Find the shapes shown in the sample below. Then, color them the same color as the sample.

Sample

A yellow triangle

What Is It?

- Use the key below to color by letter.
  r=red g=green v=violet (purple) b=black

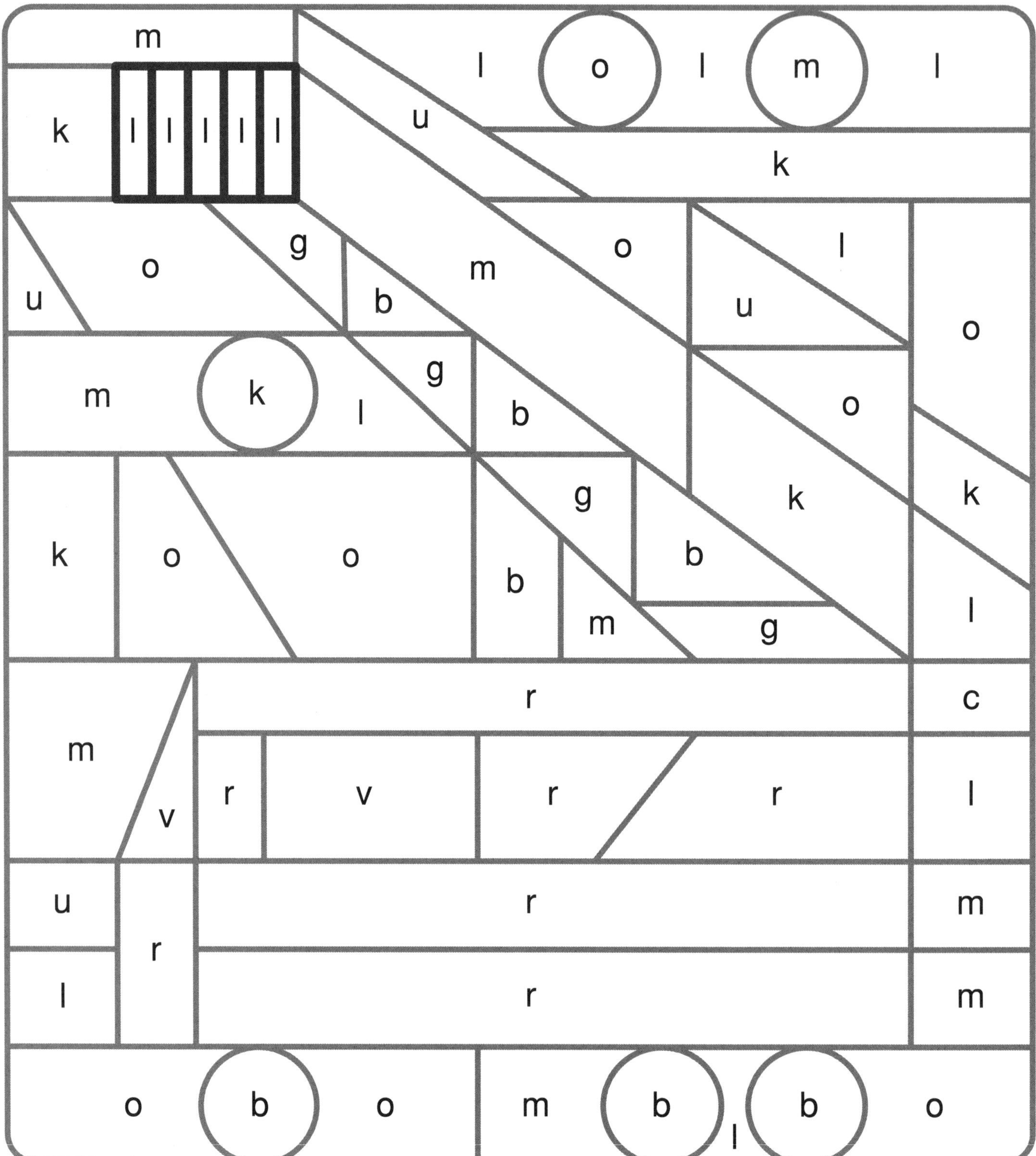

36 Which is the Same Shape and Color?

Name

Date

To parents
If your child colors the incorrect shapes or colors with the incorrect color, please tell him or her the number of the correct shapes as a hint, or mark the incorrect shapes with him or her together.

- Find the shapes shown in the sample below. Then, color them the same color as the sample.

Sample

A green square

A black hexagon

What Is It?

- Use the key below to color by letter.
  r=red g=green b=blue v=violet (purple)

37 Which is the Same Shape and Color?

Name

Date

To parents
It might be difficult for your child to find the correct shape if it is upside down. If necessary, please give him or her a hint.

■ Find the shapes shown in the sample below. Then, color them the same color as the sample.

Sample

An orange circle

A brown diamond (rhombus)

A violet (purple) trapezoid

What Is It?

Use the key below to color by letter.
o=orange   y=yellow   g=green   v=violet (purple)   b=brown

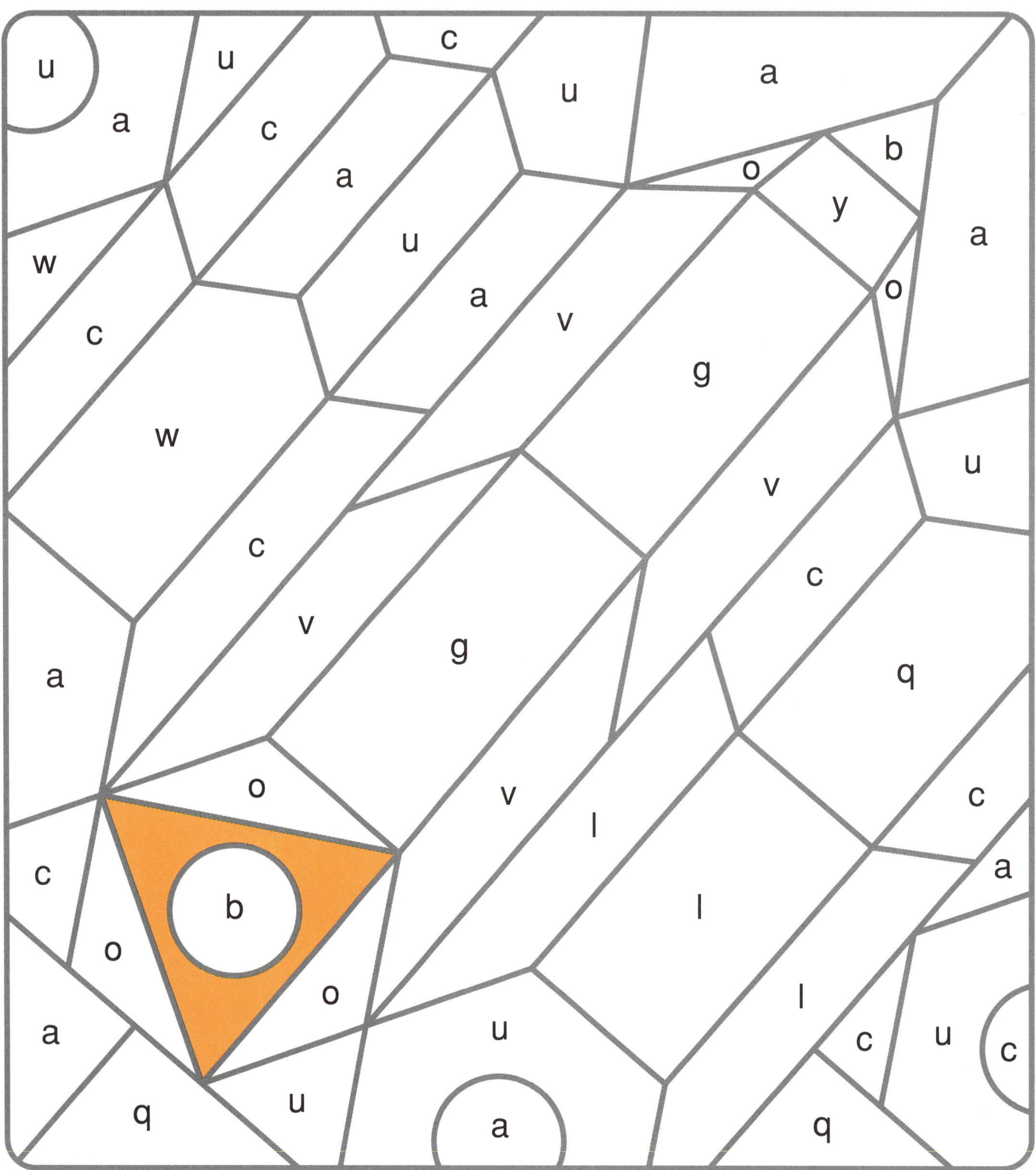

38 Which is the Same Shape and Color?

Name

Date

To parents
This is the last page of this workbook. Offer your child a lot of praise for his or her effort and accomplishment. Give your child the Certificate of Achievement on which you can write his or her name and the date.

■ Find the shapes shown in the sample below. Then, color them the same color as the sample.

Sample

A red pentagon

A blue oval

A yellow parallelogram

A green rectangle

What Is It?

■ Use the key below to color by letter.
r=red o=orange y=yellow g=green b=blue v=violet (purple)

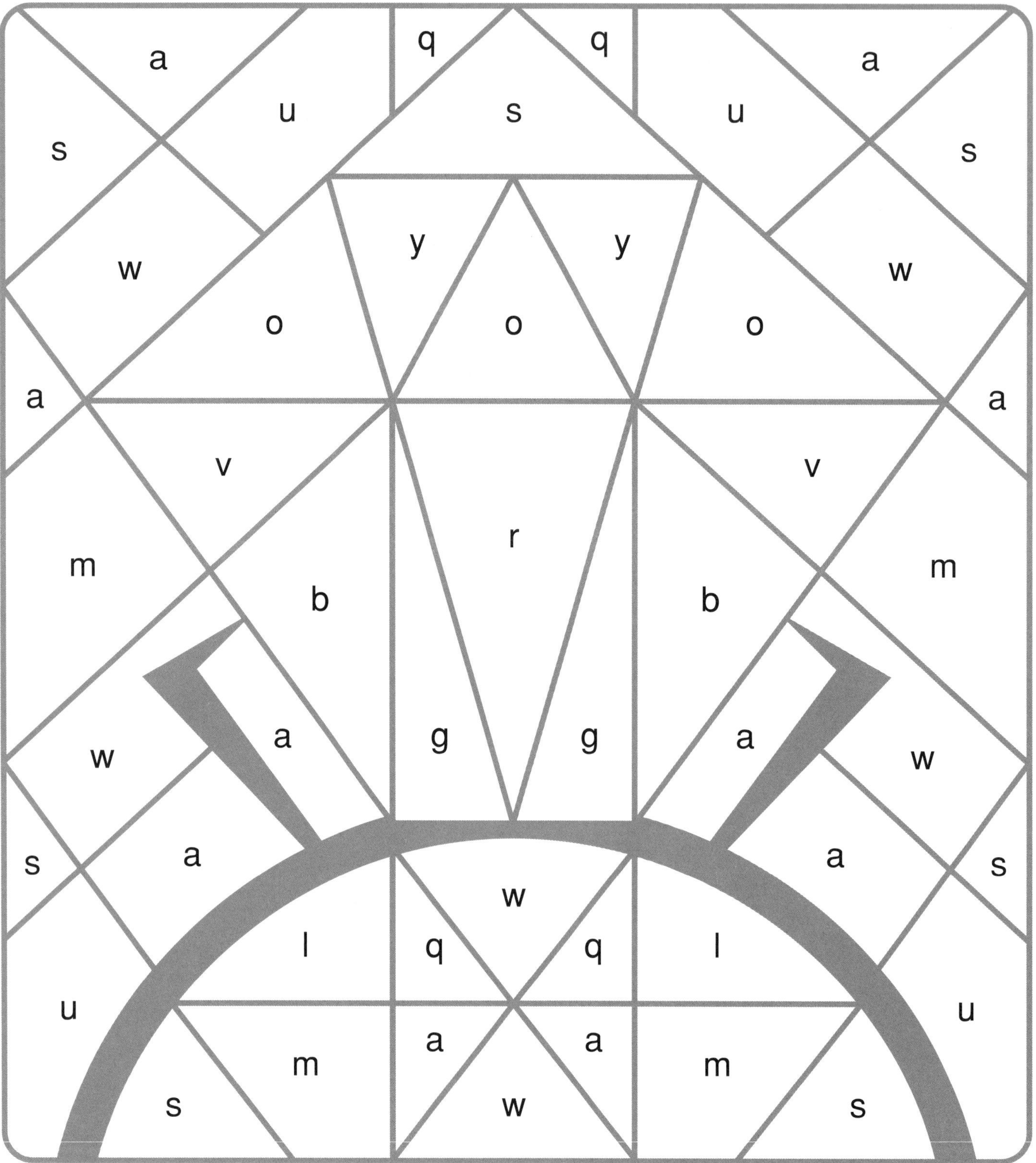

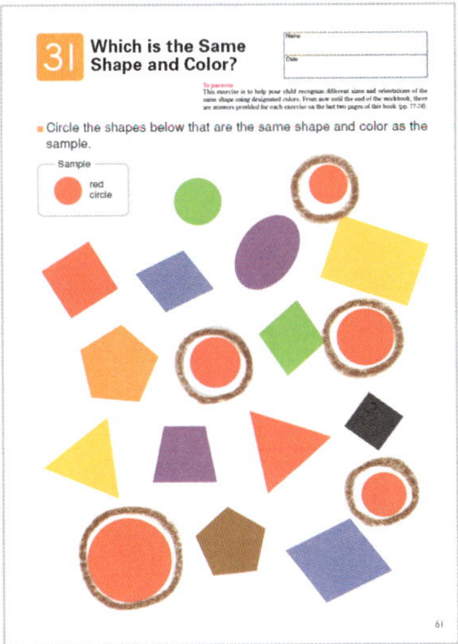
page 61

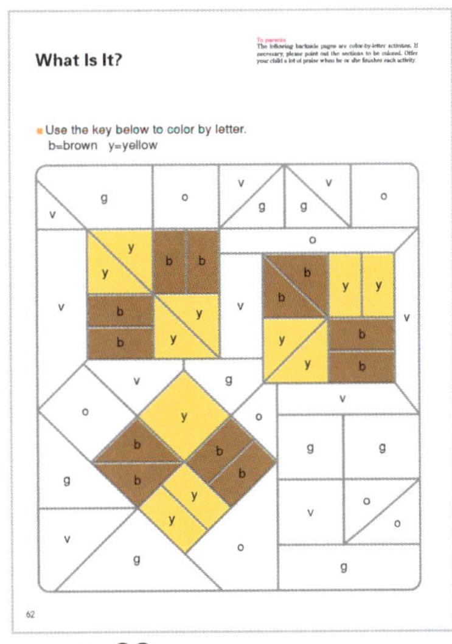
page 62
cookies

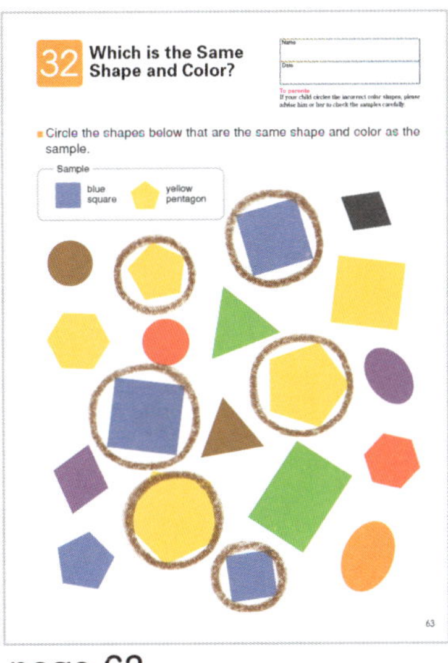
page 63

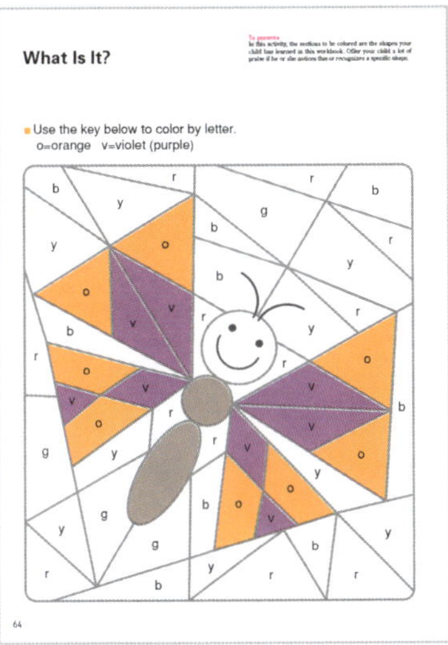
page 64
butterfly

page 65

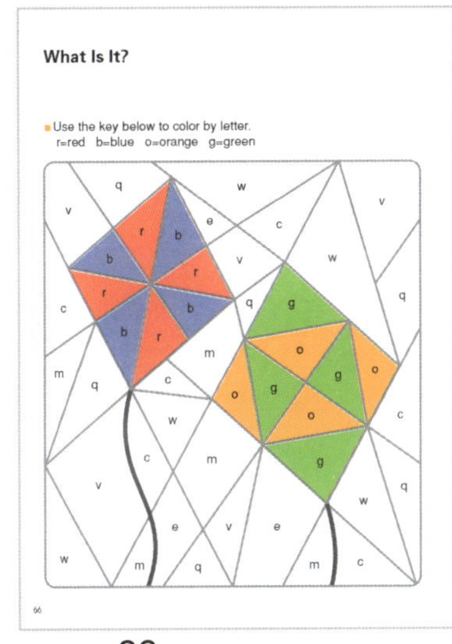
page 66
kites

page 67

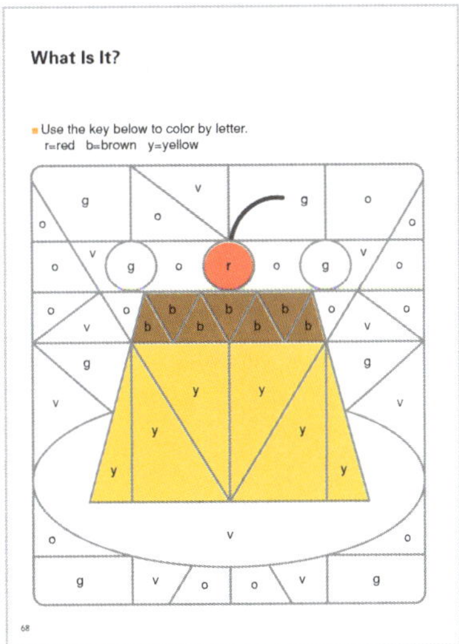
page 68
pudding and cherry

Answer Key

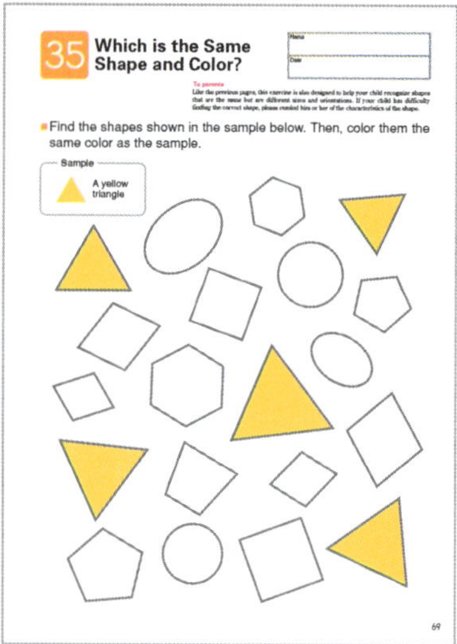
35 Which is the Same Shape and Color?

Find the shapes shown in the sample below. Then, color them the same color as the sample.

Sample: A yellow triangle

page 69

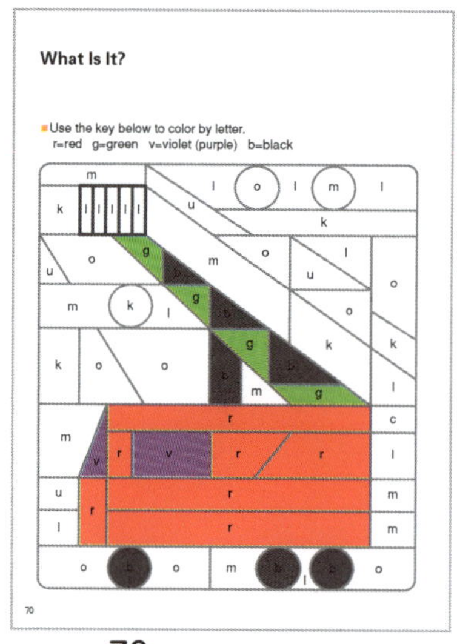
What Is It?

Use the key below to color by letter.
r=red g=green v=violet (purple) b=black

page 70
firetruck

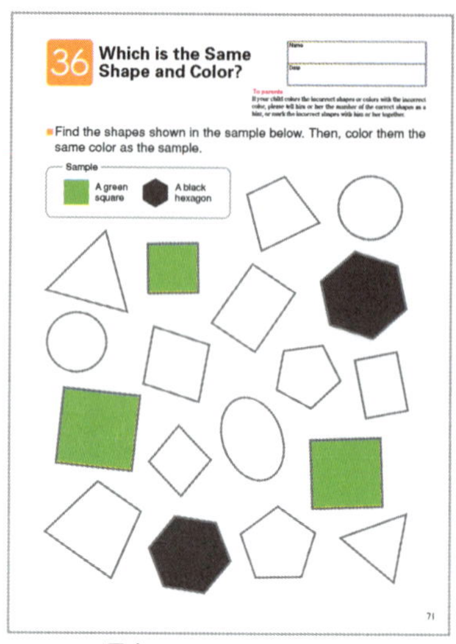
36 Which is the Same Shape and Color?

Find the shapes shown in the sample below. Then, color them the same color as the sample.

Sample: A green square; A black hexagon

page 71

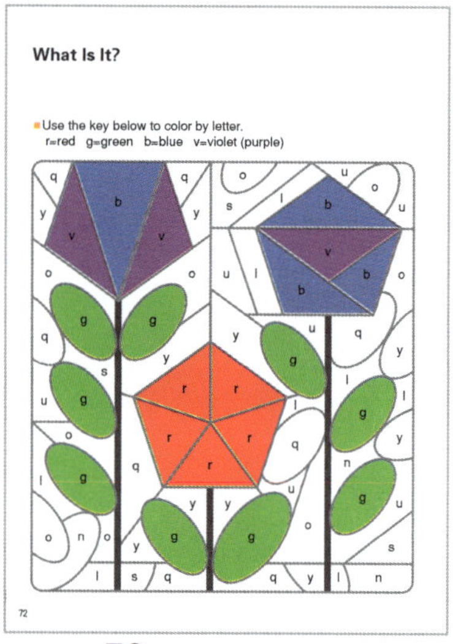
What Is It?

Use the key below to color by letter.
r=red g=green b=blue v=violet (purple)

page 72
flowers

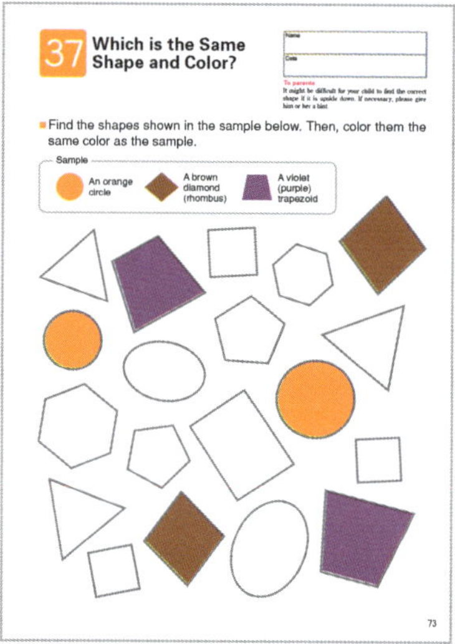
37 Which is the Same Shape and Color?

Find the shapes shown in the sample below. Then, color them the same color as the sample.

Sample: An orange circle; A brown diamond (rhombus); A violet (purple) trapezoid

page 73

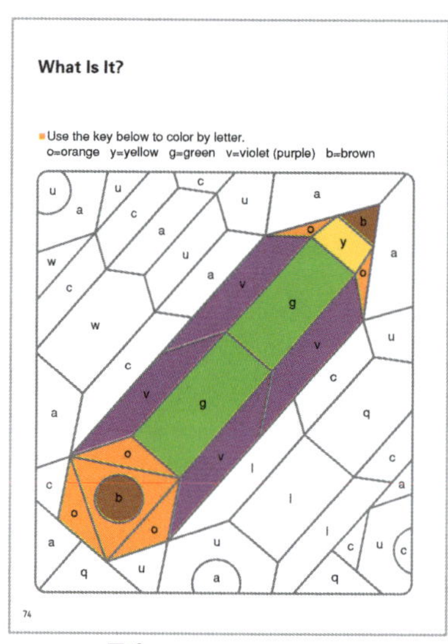
What Is It?

Use the key below to color by letter.
o=orange y=yellow g=green v=violet (purple) b=brown

page 74
pencil

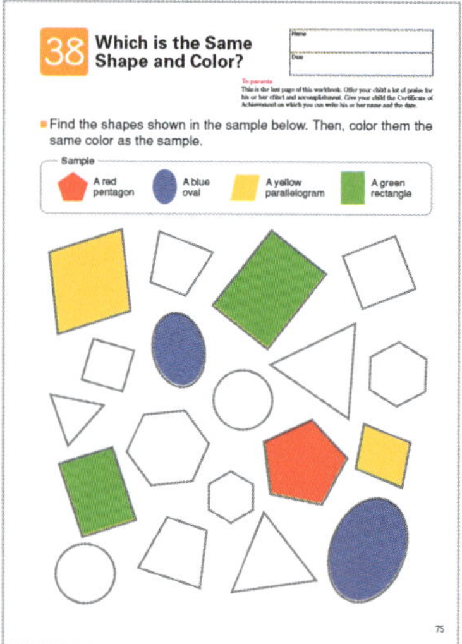
38 Which is the Same Shape and Color?

Find the shapes shown in the sample below. Then, color them the same color as the sample.

Sample: A red pentagon; A blue oval; A yellow parallelogram; A green rectangle

page 75

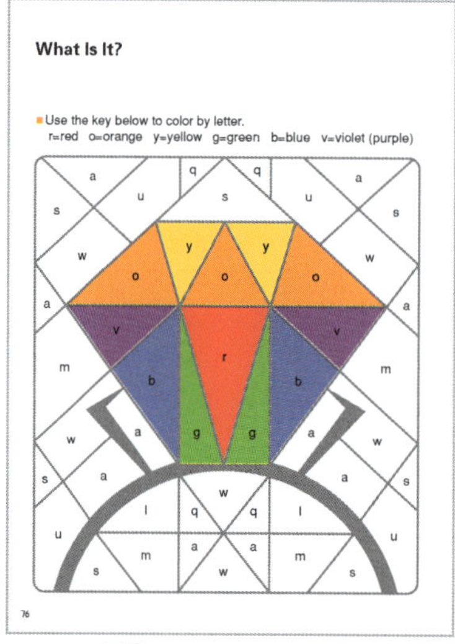
What Is It?

Use the key below to color by letter.
r=red o=orange y=yellow g=green b=blue v=violet (purple)

page 76
jewel

KUMON

Certificate of Achievement

____

is hereby congratulated on completing

My Book of Shapes & Colors

Presented on ______________________, 20___

____

Parent or Guardian